Cecilia Payne-Gap
an autobiography a

Cecilia Payne 1919 (Courtesy of the principal and fellows of Newnham College, Cambridge, England)

Cecilia Payne-Gaposchkin
an autobiography and other recollections

EDITED BY
KATHERINE HARAMUNDANIS

Second edition

Published by the Press Syndicate of the University of Cambridge
The Pitt Building, Trumpington Street, Cambridge CB2 1RP
40 West 20th Street, New York, NY 10011–4211, USA
10 Stamford Road, Oakleigh, Melbourne 3166, Australia

© Cambridge University Press 1984, 1996

First published 1984
Second edition 1996

Library of Congress catalogue card number: 83–5290

British Library cataloguing in publication data

Payne-Gaposchkin, Cecilia
Cecilia Payne-Gaposchkin: an autobiography and
other recollections
1. Payne-Gaposchkin, Cecilia   2. Astronomers—
Biography
I. Title   II. Haramundanis, Katherine
III. Payne-Gaposchkin, Cecilia. The dyer's hand.
520'.92'4   QB36.P/

Library of Congress cataloguing-in-publication data available

ISBN 0 521 48251 8 hardback
ISBN 0 521 48390 5 paperback

Transferred to digital printing 2001

UP

Contents

Cecilia Payne-Gaposchkin: an introduction *by Virginia Trimble*		vii
An introduction to 'The dyer's hand' *by Jesse L. Greenstein*		1
An historical introduction to 'The dyer's hand' *by Peggy A. Kidwell*		11
A personal recollection *by Katherine Haramundanis*		39

The dyer's hand: an autobiography
by Cecilia Payne-Gaposchkin

	Dedication	70
	Foreword	71
	PART I THE VISION SPLENDID	75
1	Backgrounds	77
2	Beginnings	84
3	Prelude to education	90
4	Birth of a dream	96
5	Dramatic interlude	104
6	The dream fulfilled	108
7	Pathway to the stars	112
	PART II THE LIGHT OF COMMON DAY	127
8	England and the United States	129
9	Harvard College Observatory	137
10	The cradle of astrophysics	144
11	Harlow Shapley	154

12	Stellar atmospheres	159
13	Spectra and luminosities	167
14	Editorial experiences	173
15	Visiting astronomers	177
16	At the cross roads	182

PART III THE DYER'S HAND SUBDUED — 187

17	Turning point	189
18	Prolegomena to variable stars	198
19	International problems	203
20	End of an era	207
21	Retrospect	213

PART IV REFLECTIONS — 217

22	On being a woman	219
23	Science and myth	228
24	Worlds not realized	234

Bibliography of works by Cecilia Payne-Gaposchkin — 239

Postlude — 256

Index — 259

Cecilia Payne-Gaposchkin: an introduction

VIRGINIA TRIMBLE

University of California and University of Maryland

My happiest memory of Cecilia Payne-Gaposchkin comes from the evening of her Russell Prize Lecture. It was January 17th, 1977, the 149th meeting of the American Astronomical Society, in Honolulu, Hawaii. Most participants had gone native, in muus and brightly colored shirts (though perhaps not quite so much so as would happen now). But not Cecilia. She wore a long, full, black gown, made of some taffeta-like fabric, overlaid with lace, in a style that had not been readily found in stores for many decades. It was easy to imagine the kind of family background ('We Paynes, my dear, do not *buy* our frocks; we *have* our frocks.') that shows through the early sections of 'The dyer's hand' and that contributed to the backbone, both literal and metaphorical, with which she faced the world. Inept as ever, what I said to her after the lecture was not 'What a wonderful talk!' but 'What a wonderful dress!' She was, at least, not obviously displeased.

That the former was also true, you can see for yourself, since the printed text (Payne-Gaposchkin, 1977) closely resembles the talk as given. I can still hear her echoing the joyous shout of Henrietta Swann Leavitt, 'That star hasn't been seen for almost 30 years!' describing the 1919 reappearance of T Pyxidis, nova 1890. The Russell Lecture was entitled 'Fifty years of novae', and CHPG (as I shall frequently call her hereafter) had indeed been looking at these enigmatic exploding stars for about half a century, with her definitive monograph (*The galactic novae*, Payne-Gaposchkin, 1957) appearing somewhat past the half-way point. The book became and remains her single most-cited publication.

But her greatest contribution to astronomy was something quite different – the demonstration that normal stars all have essentially the same chemical composition, dominated by hydrogen and helium. The data, the analysis, and the conclusions are all in the hardback version of her doctoral dissertation (Payne, 1925) as well as in several papers published at about the same time. The conclusion was, however, so unexpected that, following advice from Henry Norris Russell and Harlow Shapley (her official advisor), she attributed the remarkably strong lines of hydrogen to 'anomalous excitation' rather than to enormous abundance. The phrase means that there might be a much larger fraction of the hydrogen atoms in the $n = 2$ level, able to produce absorption lines in visible light, than would occur in thermal equilibrium.

Modern historians of science (e.g. Hufbauer, 1991 and references therein) firmly credit her with the discovery that most things in the universe are made mostly of hydrogen. Payne's more senior colleagues (e.g. Eddington, 1926) slogged on trying to understand stars as an earth-like mix of oxygen, silicon, iron, and such, until Russell, in a remarkable about-face, decided in 1929 that hydrogen was the commonest element in the Sun. Though his papers cited Payne's work as support for his result, the damage could never fully be undone; and by then Shapley was already using his directorial reins to turn her attention from stellar spectra to photometry, eclipsing binaries, and variable stars.

It is enormously tempting to ask 'what if' Dr Payne had advocated her correct conclusions a little more firmly and continued to work for the next 10 or 20 or 50 years on stellar spectra. Would she soon have recognized that a few stars had anomalously weak metal lines for their color temperatures, and so have anticipated the post-war concept of stellar populations? Might she eventually have started asking how the normal mix of elements and the deviations from it got to be the way they are and so have started astronomers on the path to understanding chemical evolution much earlier than in fact happened? Impossible, of course, to say. But she always saw her own greatest strength as lying in organizing, correlating, and systematizing large quantities of material, precisely the skills that would seem to have been needed.

The last letter I had from Cecilia was written from the Philippines on the ill-fated 'final journey' (p. 64) in the spring of 1979. It was a response to a panic-stricken request for help with a review article on binary stars in clusters. She responded generously with extracts from the accumulated wisdom that appeared later that year in *Stars and clusters*, her last book (Payne-Gaposchkin, 1979), including the tidbit that Baade had been the first to look for – and not find – eclipsing binaries in globular clusters. By the time the review appeared (Trimble, 1980), she was gone. I miss her still, in much the same way that I miss the women, older and wiser than me, of my own family.

An astronomer who happened to be a woman

The year Cecilia Helena Payne was born, 1900, the life expectancy for American white females was 51 years (and probably somewhat shorter in her native England). That she was far above average in this respect, as in so many others, is the reason that a good many current members of the astronomical community can still boast of having known her, though she started graduate work in the almost mythical past of the 'giant and dwarf theory' of stellar evolution.

Payne-Gaposchkin's last scientific collaborator (Susan Kleinmann – they identified very red stars in a two-micron sky survey) was born more than 60 years after her first (Harlow Shapley), giving a new meaning to 'a generation' of astronomers. Another aspect of this long career was her vivid memories of some important astronomical events, like the discovery of T Pyx, the first recurrent nova, just mentioned. She was not present (and could not have been) at the 1920 Curtis–Shapley debate on the distance scale of the Universe and the nature of the spiral nebulae. But she was there when the issue was resolved and has described (p. 209) Shapley's reaction to the 1924 discovery by Edwin Hubble of Cepheid variables in the Andromeda Nebula. A letter including this definitive evidence that the nebulae are separate galaxies soon reached Shapley, and he showed it to CHP saying 'This is the letter that destroyed my universe.'

Payne was not the first woman to earn a PhD in astronomy in the United States (though she was the first at Harvard/Radcliffe, closely followed by Helen Sawyer). That honor, if so it be, belongs to Margetta Palmer, who received a Yale doctorate in 1894 for a study of the orbit of Comet 1847 VI (discovered by Maria Mitchell and Mme Rümker), showing that it left the Solar System on a parabolic orbit (Hoffleit, 1991). Other earlier pioneers were Caroline Furness (PhD Columbia, 1900), Dorothea Klumpke (later Roberts, DSc Paris, 1893) and Julia May Hawkes (PhD Michigan 1920, one of several pre-1925 women doctoral recipients there, whose thesis was 'Photographic determination of the positions of stars and nebulous knots in and around the Great Nebula of Andromeda', R. L. Sears, personal communication).

Most conspicuously, perhaps, the Pickering-led years of Harvard College Observatory were filled with projects timed in 'woman years', as his bevy of (non-degree-seeking) assistants classified stellar spectra, attempted to calibrate photographic magnitude scales, and so forth. Several of these women continue to appear in the indices of modern reference books. Annie Jump Cannon, for her remarkable achievement in classifying more than 200 000 stellar spectra for the *Henry Draper Catalogue* and its extensions, Henrietta Swann Leavitt, for the discovery of the period–luminosity relations for Cepheid variables in the Magellanic Clouds, and Antonia Maury, for the discovery of the first spectroscopic binary (Mizar), are the best known. But (and the idea was expressed first and at greater length by Greenstein, 1993), C. H. Payne was pretty unambiguously the first woman to make original contributions in astronomy and astrophysics (in the sense we now expect of both genders as researchers) by inventing her own problems and then solving them.

CHPG could, therefore, quite reasonably have claimed (though she would never have done so) to be the first modern woman astronomer. As a very minor offshoot of this uniqueness, it turned out to be impossible to find a reasonable match for her in an investigation of how rates of citations to astronomical papers change with time after the death of the author (Trimble, 1986). Though the goal was a match in career length, native language, and gender, the closest I could come was Sir William H. McCrea, whose first paper appeared in 1925. Cecilia's native language, incidentally, remained British English throughout her 56 years in the US. Her account (p. 133) of being accused of this after only 20 years abroad is followed immediately by 'expect' and 'want' used in distinctly non-American ways. That time, at least, I expect she did it on purpose.

Payne-Gaposchkin's scientific record is remarkable by any standard. With papers appearing from 1923 to 1979, she does not quite equal Joel Stebbins' longevity record of 64 years (1901–64), but she does tie for fifth place with W. W. Morgan (1927–83), the intermediate places being held by Philip C. Keenan (1931–89), Walter S. Adams (1900–57), and Alfred H. Joy (1917–74); and she tops the only ranked female, Anne B. Underhill (49 years, from 1946 to 1994, though these numbers are admittedly only for publications in the *Astrophysical Journal*, Abt, 1995).

Longevity aside, very few astronomers have solved a major problem in the field as part of a PhD thesis, in the way that CHPG answered the very basic question of what stars are made of. Another who comes to mind is Beatrice Muriel Hill Tinsley (1941–81), whose dissertation (Tinsley, 1968) at the University of Texas, Austin was the first numerical treatment of evolution of galaxies. It is possible to draw a number of other analogies between them. Both were among the first to earn PhDs in astronomy at their institutions, and both completed the work in something less than two years. Both played the violin, wrote poetry, and constructed unlikely artifacts out of improbable materials in their spare time. And both were clearly undercompensated much of the time for their work as a direct result of being women. That both were enormously talented and determined is simultaneously too obvious to say and too important to leave out.

Another curious analogy – the first Annie J. Cannon Prize, for astronomical achievement by a woman, went to CHPG in 1934. Eleven winners later, the prize temporarily disappeared, when

E. Margaret Burbidge declined to accept it, on the grounds that such gender-specific awards were no longer needed or appropriate. This event was important enough to Beatrice Tinsley that she mentioned it in her letters to her family (Hill, 1986). And, when the Cannon Prize was revived under different ground rules in 1974, the first winner was – Beatrice M. Tinsley (the same age, within a year, as CHPG when she won). Though their careers were spent primarily in the United States, Cecilia, Margaret, and Beatrice were all born in England.

Payne-Gaposchkin's Russell Lecture, though the 29th in the series, was the first by a woman. She was followed by Eleanor Margaret Peachey Burbidge (1984) and Vera Cooper Rubin (1994).* Not so very long ago, whether the totality of American Astronomical Society Prize winners included 10% or 12% women depended entirely on whether you counted Cecilia and Margaret (the first woman to receive the Warner Prize for young astronomers) once each or twice.

One way of tracing CHPG's influence into the present is through citations to her books and papers. On average, at present, an American woman astronomer has the papers (etc.) of which she is sole or senior author cited about 32 times per year in papers written by other astronomers (yes, the average for men is higher, Trimble, 1993, when corrected for the improper inclusion of Charlotte Moore Sitterly), and the rate for any astronomer drops rather systematically after death (Trimble, 1986). Over the quinquennia for which data are readily available, CHPG's work has been cited at a remarkably steady rate (see Table 1).

* Margaret Burbidge (b. 1919 in Manchester) has spent most of her career looking at spectra, first of stars, later of galaxies, and currently of quasars. She was part of the quintumvirate (Burbidge, Burbidge, Fowler and Hoyle, 1957, Cameron, 1957) that discovered and presented the full pattern of abundances of the elements in stars and galaxies and identified the sets of nuclear reactions responsible for each. That both E. M. Burbidge and B. M. Tinsley are best known for contributions in the areas into which Payne-Gaposchkin's interests would probably have developed, if she had been allowed to follow her own instincts, must mean something; but I am not sure what. Vera Rubin (b. 1928 in Philadelphia) has made her mark in a different area, finding evidence for more dark matter in galaxies and for larger-scale structures in the universe (also probably dominated by dark matter) than most astronomers had expected.

Table 1 Number of citations to papers and books by
C. Payne-Gaposchkin

Years	Number of citations
1965–69	103
1970–74	134
1975–79	205
1980–84	184
1985–89	193
1990–94	c. 175

(data from *Science Citation Index*)

In the earlier years, papers reporting light curves of individual stars were popular. Later on, the most cited items are her books and review articles (e.g. 1947 on Cepheid light curves, 1963 on R Corona Borealis variability, and 1966 on variable stars in the Small Magellanic Cloud). The papers and book containing her thesis results have not been much cited in recent years, except by historians of science, and this is typical of pioneering work. The phenomenon is called 'incorporation' by workers in scientometrics, though privately I like to think of it as the second-order Mossbauer effect – after all, who these days bothers to cite Mossbauer when reporting a result found using his effect? If you would like to check the citation numbers, watch out for multiple spellings. Not only are CHPG's papers divided between Payne and Gaposchkin, but also the middle initial comes and goes, and the surname has inspired authors to some remarkably creative spelling.

Payne-Gaposchkin's opinions of other astronomers were not uniformly high (though nearly always generous), as you will see from the pages that follow. Her obituaries of older colleagues thus make interesting reading. The first she was asked to write (Payne-Gaposchkin, 1941) was that of Annie J. Cannon, in a totally typical editorial action. To this day, editors consider that a book by a woman scientist has to be reviewed by another woman! The Cannon obituary is a fair appraisal, I think, of life and works and ends by saying, 'She was the happiest person I have ever known.' I wonder whether that remained true down through the years as society came to place higher value on happiness, or at least the appearance of it!

The obituary of Otto Struve (Payne-Gaposchkin, 1963) focusses heavily on his astrophysical contributions, which were of a type she understood and admitted. Fritz Zwicky (Payne-Gaposchkin,1974) in contrast seems to have puzzled her, though she concludes by saying of his Gold Medal from the Royal Astronomical Society 'I believe that posterity will consider that he earned it.' With the increasing importance of supernovae, gravitational lensing, and dark matter (all arguably Zwicky discoveries) nearly everybody else now believes it too. I claim, at least, to have been a very early convert.

Though CHPG did very little formal teaching, she nevertheless produced an introductory textbook (Payne-Gaposchkin, 1954). The volume is notable for the quotes from Tennyson and other poets at the heads of the chapters. This practice has since grown to the status of a cliché in astronomy, but she was among the first to do it, and among the most enigmatic (I still cannot claim to understand all the choices, even with help on contexts from Bartlett's Familiar Quotations). She was certainly the first, and so far only, woman astronomer to collaborate with a daughter on a book (the revision of that text, Payne-Gaposchkin and Haramundanis, 1970). They are not quite the only mother–daughter pair in the field (Vera Rubin and Judith Young are well known, and there may be others), but it is surely rare!

A woman who happened to be an astronomer

Hercule Poirot, providing a voice for Agatha Christie Mallowan (1891–1976, and born, like CHPG, in rural England) said repeatedly that whether a girl is physically attractive is of transcendent importance in her life. You don't really have to be a detective to

An introduction

figure that one out for yourself, but I suspect it may be even more important whether the girl believes herself to be attractive. At some deep level, the young Cecilia Payne felt that she was not (pp. 190–1). Apparently at times she even wished that she had been born a man (p. 73), and you will see hints of how the issue of physical attractiveness affected her right through to the final, reflective passages (pp. 226–7), even when she chose to treat the matter as a subject for humor ('the thin end of the wedge' pp. 29, 157). A young Cecilia Payne who had thought of herself as beautiful, feminine, or pretty might still have become an astronomer, but she would have been a very different sort of scientist and person, probably with a different surname.

Two questions immediately suggest themselves: why did she feel that way? and was it true? They are not the same, and neither is very easy to answer.

Children (so the experts tell us) paint their self portraits out of materials supplied first by their parents and later by their peers. CHPG seems to have come from a home where children were not idly praised, and where most of the praise going was reserved for the one male Payne (p. 86). She grew up in a fatherless household and attended all-girls' schools and the almost-equally segregated Newnham College, Cambridge. Thus Cecilia apparently never met a young man as a casual friend or peer until she arrived in Massachusetts. The American custom of co-education immediately struck her as a good thing (p. 134).

Mores change, slowly. Margaret Burbidge, who grew up 20 years later in a household centered around an academic father, remembers no 'personal appearance' feedback there, apart from instructions to wash her hands or change for tea. But her summer holiday excursions were less circumscribed than CHPG's and her university (London) was co-educational, so that somewhere between the ages of 15 and 17, she gradually caught on to the idea that she was attractive and that this was no bad thing to be.

Yet another 20 years downstream, Beatrice Hill Tinsley's family thought she was pleasant to look at and said so (Hill, 1986, p. 13). Though her secondary school was also women only, she had broken her first engagement at an age when CHPG had met no men except father, brother, vicars, and Gustave Holst. My own experience (in case anyone is interested) cannot be put into parallel format. I am a graduate of Hollywood High School, where no one laughs at a joke (they say, in maximal praise, 'that's funny'), and the photographer says 'chin up and to the left two inches' even, or perhaps especially, if the photographer is your father.

What was the reality? My own memories are (again!) useless. Past 50, women (and men) have the faces they deserve. The Cecilia I knew was warm and jolly. It shows in Fig. 14. So, for that matter, was Agatha Christie late in life, if her book jackets and self-portrait as Ariadne Oliver are to be trusted. Katherine Haramundanis remembers her mother as handsome in midlife (p. 39–40). Her husband described her, variously, as elegant, stately, and 'Brunhilda' (Gaposhkin, 1974, 1976, 1978). Other memories are less pleasant. A recent letter from my thesis advisor recalls an incident where a group of astronomers were discussing a suitable gift for CHPG on some significant birthday, and the winner (at least on the laugh meter) was 'a shaving machine' (Münch, 1995). This story clearly says more about the other participants than about CHPG, but it provides a glimpse of the environment in which she worked much of the time. It is by no means the most unpleasant one I heard in preparing this introduction.

The photographic record is singularly unhelpful. Compare Figs. 4 and 5, if you did not immediately skip to the pictures the instant this topic arose. It is also very sparse. Neither the American Institute of Physics historical collection nor that of the Astronomical Society of the Pacific (which I raid frequently for illustrations to review articles) has a Payne-Gaposchkin photo. CHPG wrote the second of (to date) 18 personal memoires chapters for *Annual Reviews of Astronomy and Astrophysics* (Payne-Gaposhkin, 1978). It is the only one so far not accompanied by a picture of the author. Entitled 'The Development of Our Knowledge of Variable Stars', it is also just about the most impersonal of the articles in the series, with the possible exception of that contributed by William McCrea (1987). CHPG was, of course, the first woman astronomer represented in the series. The second was (you are entitled to only one guess) Margaret Burbidge

An introduction

(1995). Her article is a fascinating mingling of life and works, preceded by a charming photograph. That there are so few photographs of CHPG to be had leads one to suspect that she felt they were not worth having.

Objectively, Payne-Gaposchkin was tall (about 5'10" or 175 cm) and broad shouldered in an era that valued petiteness. Presumably she would have wished it otherwise. But it's an ill wind that blows no good. Down to the present, male astronomers typically take more seriously women who can look them in the eye. One colleague has never quite recovered from the reaction, 'Oh, isn't it cute; the little girl has an idea', when she first ventured out of her home institution. At any rate, no one ever accused CHPG of cuteness. And I think my own path was easier, if more heavily trod, for my being some inches taller than the female average.

What about Tinsley? Photographs (Hill, 1986) confirm the early charm recorded by her mother, but, by her mid 30s, she was quite utilitarian, justifying the ruthless confinement of rather spectacular hair with the claim that 'the average wind speed in Wellington is 40 miles per hour'. (Her school and college years were spent in New Zealand.)

Nobody can ever know the whole truth about a marriage, not even her own. But in one respect, at least, Cecilia Payne was average. She married another astronomer. So did a majority of the Cannon Prize winners who followed her under the original system (a good many never married). Even now, according to a survey of the American Astronomical Society, roughly half of the married women members have spouses in the field, and the majority of the remainder have spouses in related fields like physics and the geosciences.

Sergei Illarionovich Gaposhkin (1898–1984) outlived his wife by only a few years. In contrast to her single, slim autobiographic volume, he left three massive ones (Gaposhkin, 1974, 1976, 1978), adding up to no fewer than 2002 pages. They are unindexed and arranged according to a scheme that only the author understood; so if you find a passage you like, you had better write down the page number, or you will never find it again!

To begin with, a word about names is required. Cecilia changed her surname upon marriage (and was, in fact, less inclined to hyphenate than those of us who write about her). It was the custom of the time and place. But she had thought about the matter, later, if not in 1934, and came to describe the change as a token of 'gratitude and loyalty' (p. 131). There were, however, limits. Some time in the late 1960s, Sergei dropped the (in English, at least) superfluous 'c', compactifying from Gaposchkin to Gaposhkin. He devoted an AAS contributed talk to explaining why, on the same day as CHPG's Russell Lecture (though the published abstract deals with the nature of WX Ceti, Gaposhkin, 1977). I have notes from the session at which he spoke, but still don't understand the reason for the change, having thought at the time he was threatening to switch from Sergei to Sirgay, as indeed he then did, in an attempt to make spelling agree with pronunciation for Americans. In any case, Cecilia did not follow the example and retained the earlier spelling. This adds considerably to the confusion of any bibliographic or citation search for the pair!

Among other astronomers who have been turning up in these paragraphs, Margaret Burbidge shares the surname of her astronomer husband Geoffrey, and Vera Rubin that of physicist Robert. Beatrice Tinsley was, as has become common, left with space scientist Brian's surname after their divorce. I was quickly teased out of any tendency to try to merge names with physicist Joe Weber after our marriage by a colleague who produced a few deliberate garbles like Treber-Wimble on envelopes addressed to me. A quick glance at AAS directory pages suggests that the current pattern is about 50% of astronomical marriages leave names unchanged, 30% result in both people using his surname, and 20% produce other combinations (both take hers; both hyphenate; she hyphenates, but he doesn't, etc.)

The 'gratitude and loyalty' quote (which I find distressing in the context of marriage) is actually part of a musing on naturalization. Cecilia became an American citizen just about as soon as the rules allowed, in a deliberate decision to participate in the public life of the country that had allowed her a scientific life, as England would not. Sergei did the same. Fritz Zwicky (1898–1974), whom you met a couple of pages back, because CHPG wrote an obituary of him,

An introduction

took a firmly opposite stand, declining to become an American citizen on the grounds that 'a naturalized citizen is always a second class citizen'. He returned periodically to his home canton of Glarus, Switzerland to vote. Beatrice Tinsley also never carried an American passport. Margaret Burbidge, in an interesting compromise, took out US citizenship papers after her election as president of the American Astronomical Society. (Yes, of course she was the first woman to hold the post.) Vera Rubin and I were born in the United States, though you never have to go back many generations among Americans to find the ancestor who came from 'the old country' – one generation back to Latvia for her, three to Denmark for me.

Sergei's version of his first meeting with Miss Payne (Gaposhkin, 1976, p. 1637) very much resembles hers (p. 196), though, mysteriously, he calls her 'Miss H'. He describes her as 'like a ripe peach left alone on the tree, darkened, wrinkled a little outside, but the more delicious inside' and as sexually totally inexperienced.

While she dedicated her last book (Payne-Gaposchkin, 1979) 'to my husband, "that bright particular star,"'* he scattered her image and name through many of his 2002 pages, in English, Russian, and at least once in ancient Egyptian hieroglyphics. He translates the griffitus (Gaposhkin, 1976, p. 1455) as 'love by the beautiful lady Cecilia; the great man Sergei, given life'. The transliterated names are actually Chichi(rw, used in Ptolemaic times for l)ia and Sirg(hard)y. 'Given life' was the customary attribute of a reigning pharoah. The glyph for 'love' is missing both its phonetic complements and its determinative, and the wrong preposition has been

---

* The quotation is from Act I of Shakespeare's *All's Well that Ends Well*, in which Helena describes her feelings for Bertram as '...That I should love a bright particular star, And think to wed it, he is so high above me.' Since the play is a comedy, it naturally ends with them together (under happier circumstances than, say, Romeo and Juliet in Act V). The clause 'Nature never did betray the heart that loved her' (p. 238) is by Wordsworth, from 'Lines Composed a Few Miles Above Tintern Abbey'. We assume 'above' to have meant some direction other than the purely vertical. 'The Dyer's Hand' is Shakespeare again, Sonnet 111, 'My nature is subdu'd to what it works in, like the dyer's hand.' The rest you must look up for yourself.

used so that a malicious amateur Egyptologist might claim that the group really says 'milk jug belonging to the beautiful woman'. The names and images of other women are scattered through the three volumes of *Sergei* with equal profligacy. This is presumably an aspect of what their daughter describes as his flamboyance (p. 64). Even my own copy of Volume 2 is inscribed 'To lovely Virginia. In that first moment you were the (kindred) perfection of Beauty and Kindness. For ever Yours Sergei 1978.' The word 'kindred' or 'kindred spirit' appears often in the volumes (and represents something we all seek). Two other common phrases 'curly eyes' and 'cosmic click' remain something of a mystery, though both are clearly meant as compliments. I have not found either applied unambiguously to Cecilia, but the proverbial haystacked needle would be easy in comparison.

The last word

This has to belong to CHPG, in the form of some memorable quotes.

1. The 1920 Eddington talk about which 'I felt a stir of surprise when I read an account of this same lecture in James Hilton's *Random harvest*.' (pp. 117–18). No wonder she was surprised. For her, the lecture had been an extremely personal, intellectual experience. For Charles Ranier, it was something to which the appropriate reaction was along the lines of a practical joke or student prank (though his mathematician friend, Pal, reacted a bit the way Cecilia did and is perhaps meant as a portrait of Ramanujan). My surprise comes from the discovery that Cecilia's taste in reading matter extended down to the 1940s equivalent of airport book shop paperbacks. Hilton's best-known stories are *Lost horizon* and *Goodbye Mr. Chips*. They and *Random harvest* were all filmed with Ronald Colman in the lead.

2. 'I learned, too, that nobody loves an editor' (p. 176). How true. Just wait till you have to try it!

3. 'I will not accept the conclusions of another astronomer simply because I am fond of him, or reject them because I dislike him (though I admit there is a temptation here)' (p. 227). This occurs

in close enough juxtaposition to Shapley's plaintive, 'I believed van Maanen's results ... after all, he was my friend' (p. 209) to suggest it was a lesson she had learned at least partly from Shapley's experience. Since van Maanen's result (on the supposed rotation of spiral nebulae in the plane of the sky) made it impossible for the spirals to be separate galaxies, believing him was a fatal mistake. I think my own weakness lies more in the direction of believing friends than the opposite. And 'fond' as used here is another Briticism.

4. 'Young people, especially young women ... undertake it [a scientific career] only if nothing else will satisfy you; for nothing else is probably what you will receive.' My generation came through the system from BA to PhD to postdoc to permanent position at a time when this seemed an idle warning. Society appeared to want scientists, even astronomers, and even women astronomers, and seemed amenable to rewarding them with both moderate riches and honors. With the return to normal conditions, I have come to accept the wisdom of Cecilia's view and to believe that the only honorable form of affirmative action in the context of 'encouraging young people to go into science' is to make sure that the obstacles look just as large to white males as to anyone else.

Acknowledgements

I am most grateful to astronomers Susan Kleinmann, Judith Young, Richard Larson, and, especially, Margaret Burbidge for providing information I could not easily have found elsewhere, and to Kathy Gaposchkin Haramundanis for the invitation to write this introduction.

May 1995

Abt, H. A. 1995. Preprint
Burbidge, E. M. 1995. *Annual Reviews of Astronomy and Astrophysics*, **32**, 1
Burbidge, E. M., G. R. Burbidge, W. A. Fowler & F. Hoyle 1957. *Reviews of Modern Physics*, **29**, 547

Cameron, A. G. W. 1957. *Publications of the Astronomical Society of the Pacific*, **69**, 201; Chalk River Report CRL-41

Eddington, A. S. 1926. *The internal constitution of the stars*, Cambridge

Gaposhkin, S. I. 1974, 1976, 1978. *Sergei*, Vols. 1, 2 and 3. Privately printed

Gaposhkin, S. I. 1977. *Bulletin of the American Astronomical Society*, **8**, 520

Greenstein, G. S. 1993. The Ladies of Observatory Hill, *American Scholar*, **62**, 437

Hill, E. O. 1986. *My daughter Beatrice*. New York: American Institute of Physics

Hoffleit, D. 1991. *Astronomy at Yale, 1701–1968*, Memoires of the Connecticut Academy of Sciences, v. 23, p. 92

Hufbauer, K. 1991. *Exploring the Sun*, Baltimore: Johns Hopkins University Press, p. 102–3 and references therein

McCrea, W. H. 1987. *Annual Reviews of Astronomy and Astrophysics*, **25**, 1

Münch, G. 1995. Letter of 3 April 1995 to VT

Payne, C. H. 1925. Stellar atmospheres. Cambridge: Heffer and Sons

Payne-Gaposchkin, C. H. 1941. *Science*, **93**, 443

Payne-Gaposchkin, C. H. 1954. *Introduction to Astronomy*, Prentice-Hall

Payne-Gaposchkin, C. H. 1957. *The galactic novae*. New York: Interscience

Payne-Gaposchkin, C. H. 1963. *Sky and Telescope*, **25**, 308

Payne-Gaposchkin, C. H. 1974. *Sky and Telescope*, **47**, 311

Payne-Gaposchkin, C. H. 1977. *Astronomical Journal*, **82**, 665

Payne-Gaposchkin, C. H. 1978. *Annual Reviews of Astronomy and Astrophysics*, **16**, 1

Payne-Gaposchkin, C. H. 1979. *Stars and clusters*. Cambridge: Harvard University Press

Payne-Gaposchkin, C. H. and K. Haramundanis 1970. *Introduction to astronomy*, 2nd edition, Prentice-Hall Inc.

Tinsley, B. M. 1968. *Astrophysical Journal*, **151**, 547

Trimble, V. 1980. in J. E. Hesser, Ed, *Star Clusters*, IAU Symposium 85. Dordrecht: D. Reidel, p. 259

Trimble, V. 1986. *Journal of Czechoslovakian Physics*, **B36**, 175

Trimble, V. 1993. *Quarterly Journal of the Royal Astronomical Society*, **34**, 235

An introduction to 'The dyer's hand'

JESSE L. GREENSTEIN
Emeritus Professor of Astrophysics, California Institute of Technology, Palomar Observatory

If the scheme of philosophy which we now rear on the scientific advances of Einstein, Bohr, Rutherford and others is doomed to fail in the next thirty years, it is not to be laid to their charge that we have gone astray. Like the systems of Euclid, of Ptolemy, of Newton, which have served their turn, so the systems of Einstein and Heisenberg may give way to some fuller realisation of the world. But in each revolution of scientific thought new words are set to the old music, and that which has gone before is not destroyed but refocussed. Amid all our faulty attempts at expression the kernel of scientific truth steadily grows; and of this truth it may be said – the more it changes, the more it remains the same thing.
(from A. S. Eddington, *The nature of the physical world*, p. 353, Cambridge: Cambridge University Press, 1928)

Can we know of what stars and the Earth are made?

Sir Arthur Eddington's marvellous book starts with a comparison of two tables. There seems to be a substantial one on which he writes with a substantial pen and a mostly empty, scientific table and pen. In the scientific 'real' table, molecules, atoms, and electrons madly rush about, providing the illusion and practical effect of solidity, but occupying an infinitesimal fraction of space. The forces that provide an illusion of solidity are electromagnetic. The table is transparent to short-wavelength photons (X-rays or gamma rays). A more subtle particle, the neutrino, penetrates the table, reader, writer and Earth. The linkage between the two worlds, the familiar and the scientific, has since become even more remote; physicists would now object to ascribing energy, size and location to the individual electrons, protons and neutrons of which 'matter' is constituted. A last reality for which we can hope is that the pen deposits ink on paper in a form that communicates, and that the ideas communicated relate truly both to the illusory world of the senses and to the mad dance of forces. Quantum electrodynamics and chromodynamics are accepted, and grand unification theory, in which relativistic gravitation and particle physics will ultimately be expressible, may still arise. But this all gives little promise of restoring simple hard reality. The reality is in the process of attempting to interact or explain.

The quandary is not new. The limited complexity of the world faced by Greek philosophers and geometers was explained by imagining that real things were composites of four elements, earth, air, fire and water (pre-Socratic); but the Pythagoreans felt that numbers and their mystic relations represented the essence of things. The idea that all matter was made of similar non-divisible atoms came with Democritus. Both Pythagoras and Democritus were partly right – number and atoms are basic to the explanation of the physical world which began in the sixteenth century. Of the major forces, Newton's gravitation was the first to lead to successful prediction, with improvement by Einstein. A heritage of philosophy and of anthropocentrism left human reason with many questions viewed as being beyond physics. It was easy to create boundaries to human knowledge. These were often set by misunderstanding of philosophy and religion. My epigraph, the last phrases of Eddington, are those of a brilliant and religious man, who played a most decisive role in the early life and education of Cecilia Payne.

'The dyer's hand', a brief personal record she left of her youth and scientific career, covers a period when knowledge of the world of atoms moved from Democritus to that of Rutherford, Bohr, Einstein and Heisenberg. What atomic theory she applied to astronomy was being developed as she used it. Youth, the fact that she was a woman, that she was educated in Cambridge, England and moved to Cambridge, Massachusetts, for her most productive years, combined to make her achievements most remarkable. She could not have used quantum mechanics – it was being developed as she worked on the composition of the stars. She could not see the connection between stellar composition and nuclear energy generation in stars, because only rudimentary knowledge existed concerning the existence of nuclei and of nuclear reactions.

What she accomplished by 1925 represents an essential step in the scientific demonstration of a philosophical concept: that natural bodies, the stars, the Sun and the Earth are made of the same stuff. Atoms of different kinds have similar ratios of abundances everywhere (except for nearly obvious and important exceptions). While it remains unclear what matter is, it is manifest to us as a variety of chemical compounds of about 90 stable chemical elements. If one were creating a Universe why was the choice to create 90 stable elements? In fact, 1000 isotopes (elements with the same number of nuclear charges, but different atomic weights) exist, many of them unstable to fast or slow radioactive decay. For a classical philosopher, four elements, or one, would have been preferable, but nature, as observed near the end of the nineteenth century, is complex. Evidence on this complexity came from various sciences. Samples of rocks, at the surface of the Earth, had complex individual histories and differed from each other radically in their constituent elements. Lava from the interior had various compositions. The geologists and geochemists provided an important clue to the heterogeneous nature of rocks and, by grouping fundamental rock types, could understand why certain elements were commonly associated, in relatively constant proportions, but chemically differentiated. In ore bodies, gold came with silver; most metallic elements that could combine with oxygen (dominant in our atmosphere) had done so, forming oxides. Chemically active gases combined with reactive metals; water was a common part of the crystals. Many surface rocks were highly chemically processed and evolved, but commonly contained iron, silicon, magnesium and aluminum, in compounds that locked in whatever hydrogen and oxygen they could hold according to the laws of chemistry. This slag-heap of Earth, a mixture of relatively heavy, intrinsically complex atoms, has little to do with what might be expected of this primitive 'hyle', a word used by philosophers for matter. Aristotle called it ὕλη (wood or matter in Greek); 2000 years later George Gamow revived it as ylem. Scientists view matter as interchangeable with energy. In the 'Big Bang' there was initially only energy, soon precipitating as neutrons which in a few thousand seconds decayed to hydrogen. The complexities of composition of what we call matter are late arrivals, in current belief, created in the nuclear furnaces and explosions of stars. Just as Aristotle saw it, primitive, simple, matter–energy could receive form. We have localized and made detailed studies of that process of receiving form.

Perhaps the particular details of the untidiness of the Earth's composition could have been local accidents; the Earth is a negligible sample of the Universe. H. N. Russell (1914) had introduced the idea of a universal composition. The only matter we handle from outside the Earth is that in meteorites; co-workers at the U.S. Geological Survey, F. W. Clarke and H. S. Washington began in 1889 to study the 'relative abundance of the chemical elements', hoping to see some sort of periodicity in the abundances connected to that of the periodic table of the chemical elements. No simple regularity appeared, except that elements with even atomic numbers were more abundant than those with odd (Harkins' rule), a numerology which would have pleased the Pythagoreans but, in fact, arises from the greater stability of the even-numbered nuclei. The collection and study of stones that fall from the sky as meteorites has been an active pursuit. Their compositions are more nearly unaffected by billions of years than are the surface rocks of the Earth, subject to geological history.

By 1937, V. M. Goldschmidt had made it possible to link together minerals of very different chemical natures to produce a regular pattern of abundances of elements and isotopes. Goldschmidt was a pioneer in the modern science of geochemistry; in his classic book *Geochemistry* (Oxford: Clarendon Press, 1954), published posthumously, he quotes extensively from Cecilia's work on the cosmic abundance of hydrogen which made Jupiter more like the stars Cecilia analyzed than like the Earth. The classic results of geochemical analysis were presented by H. E. Suess and H. C. Urey (1956). They gave the relative proportions of 83 elements and 300 isotopes; one type, the so-called chondritic, stony meteorites seemed to provide a natural 'best' average over the complex composition and chemical fractionation of the Earth and planets. With such an assumption, the Earth and meteorites proved very similar; furthermore, the heavy-element composition (omitting volatile gases) of the surface layers of the Sun as studied by H. N. Russell (1929) was found to resemble that of the Earth. One large difference in composition exists within our system of planets, however, and made these only partial triumphs. The Earth and Mercury had a mean density 5.5 times that of water; the giant planets, Jupiter and Saturn, had densities 1.3 and 0.7 times that of water. Dense rocks would be even further compressed, more dense, if Jupiter were made of Earth-stuff. Instead, it is obviously mostly hydrogen and helium, as are the Sun and stars. The 'terrestrial planets' have lost their volatile gases, hydrogen and helium, and many other chemically inert gases. If all the solar hydrogen and helium were to be removed, the residue would have only 1 % of its present mass; similarly, Jupiter, at present 317 times the mass of Earth, would be only 3–5 Earth masses if only elements forming solids, like carbon, silicon and the metals, were left. Fortunately for our existence, the Earth had too small a mass to be a giant gaseous planet, and its rocks and atmosphere provide the only suitable habitat for life that we know. Knowledge of geochemistry and studies of planetary atmospheres and surfaces, and meteorites combine now with nuclear astrophysics to give a plausible model of the origin of the chemical elements in the Earth, Sun and planets; Cecilia's contribution was to make clear the nearly constant composition of the stars.

The attempt to evaluate the fractional abundance of hydrogen in the Earth, e.g., by Clarke and Washington (1922), was doomed by the volatility of hydrogen, its escape from our atmosphere, its presence only as chemically bound. Thus, a 1922 terrestrial value was 15 %, where oxygen was 55 %, silicon and iron 18 %, by number of atoms. The meteorites hardly improved the situation. So strong was this evidence that in *Stellar atmospheres* (1925: pp. 188–9), Miss Payne was forced to say

> ... In the stellar atmosphere and the meteorite the agreement is good for all atoms that are common to the two, but several important elements are not recorded in the meteorite. The outstanding discrepancy between the astrophysical and terrestrial abundances are displayed for hydrogen and helium. The enormous abundance derived for these elements in the stellar atmosphere is almost certainly not real. Probably the result may be considered, for hydrogen, as another aspect of its abnormal behavior ... The lines of atoms appear to be far more persistent, at high and low temperatures, than those of any other element.

How close to the truth she was is remarkable – hydrogen and helium make up 98 % of the matter of the Universe. How she came to doubt her own conclusions is another complex study in the history of major discoveries in science. I have found on several major occasions, myself, that excess of knowledge, and ingrained conservatism, have buried into my subconscious an obviously correct, but radical, conclusion. Where she faced a separate and easier question, an apparent difference in high hydrogen abundance in the hot stars, low in the Sun, she was in conflict with the leading astrophysicist of those years, Russell. It is interesting that 3 years later, W. S. Adams and H. N. Russell (1928) analyzed a group of seven cool supergiants, and giants, and found them roughly similar to the Sun; but they note that: 'The great intensity of the hydrogen lines in the cooler giant stars, and its conspicuous dependence on absolute magnitude have long been a puzzle. Hydrogen, being light, is peculiarly subject to radiation pressure... [but] other factors may be at work.' One is left with the strong feeling of excellent minds working at too early a date for the solution to be forced on the scientist. By 1929 Russell notes that Payne's hydrogen abundance in stars is 25 times his solar value, lists other difficulties, and shows they might be removed if 'in other words, the solar atmosphere really does consist mainly of hydrogen' and points out that the 1928 Adams and Russell analyses of stars would also agree with Payne's. To quote: 'If this is true, the outer portions of these stars must be almost pure hydrogen, with hardly more than a smell of metallic vapors in it.'

Two books by Cecilia H. Payne

Cecilia graduated from Newnham College, Cambridge, England, in 1923 and came to the Harvard College Observatory. Her PhD thesis at Radcliffe College was published in 1925 as '*Stellar atmospheres*', a book of 215 pages, filled with the novelty of atomic and spectroscopic physics, interpreting the results of measurements of available stellar spectra to give temperatures, pressures and composition for a wide variety of stars. It ends with a comparison of stellar, terrestrial and solar abundances of many elements. The treatment, by statistical mechanics, of atomic energy states had been developed by R. H.

Fowler (1923). The chemical laws of mass action had been generalized to give a theory of ionization by M. N. Saha (1921); R. H. Fowler and E. A. Milne (1923, 1924) had given a more detailed set of equations. Cecilia's book is filled with references to current and unpublished work; it was written during the great explosion of understanding of the atom. Bohr's classification of the electron orbits responsible for atomic spectra was published only in 1923. The incredible timeliness of her work strikes the reader, nearly 60 years later. She read everything applicable and, long before the atomic data were available or reliable, she used (or even corrected) all that she read. A typical scientific paper merges with the historical background in seven years. Her book is now seldom referred to; what it contains is part of our daily heritage as scientists. It is stimulating, even now, because it often faced puzzles not then, or yet, explicable. She notes the existence of stars with apparently abnormally high abundances of individual elements, like silicon, strontium and the rare earths. A rich memory, an eye for detail, are accompanied by bravery in applying the then new physics to the stars. The astronomers at other observatories had far better spectroscopic facilities than did Harvard, where she had to rely on a collection of 'objective-prism' plates of low resolution, and a few more detailed spectra of bright stars. The Dominion Astrophysical Observatory (Victoria, B.C., Canada), Mount Wilson (Pasadena, CA), and Lick Observatory (Mt. Hamilton, CA) were sources of published data for her analysis. One colleague, Donald H. Menzel, a pioneering astrophysicist in the USA, had come from Lick with a thesis on the emission-line spectrum of the solar chromosphere (seen during eclipse). Cecilia and Donald collaborated briefly; he turned to more sophisticated and mathematical treatments of the emission from hot, ionized gases (emission nebulae) and became a spiritual leader to many of my generation of astrophysicists.

Cecilia includes a brief description of the classification of stellar spectra, based on the extraordinary work of Annie J. Cannon. The 200 000 stars of the *Henry Draper Catalogue* were classified on low-resolution spectra covering the entire sky. No quantitative measurements were possible; a brief glance may decide, for us, whether an animal is elephant or bear. Miss Cannon could separate the rogue from the good elephant, or the grizzly from the brown bear, at a glance. There are about 10 'types' measured by the Draper class; each class is further subdivided. The hottest stars show He^+ lines, progressively cooler stars He, H, ionized and neutral metals, while the coolest have molecular bands. On the surface, each Draper subtype apparently had a different composition; Miss Payne (although not the first) recognized that the main variable is the degree of thermal ionization, with temperature dropping from 35000K for the O stars, to 3000K for the M stars. The classification scheme was empirical in nature, like the botanists' classification of plants (before evolution was understood). The unifying theme for stellar classification is constant composition and varying temperature, but an evolutionary theme also existed.

I met Cecilia in 1927, when I was 18, a junior undergraduate at Harvard, but only 9 years younger than she. *Stellar atmospheres* had appeared and she was involved in the work that led to her second book *The stars of high luminosity* (1930). For me, she was a quite extraordinary figure; broadly informed in astronomy, of imposing stature and stormy personality, and widely read in current English and American literature and classical music. I was seriously uncertain whether I wanted to be a poet, critic, future scientist or businessman. Cecilia could quote (and without error) Gilbert and Sullivan, T. S. Eliot or Latin epigrams. I shared her enthusiasm for the latter two; it was not uncommon to find her personally upset or tense, but willing to talk for hours about literature, or science. I enormously enjoyed those hours; although I never worked with her on a scientific topic I received her important insight that the particularities of a scientific question should never be glossed over. Perhaps I knew more current theory of atomic physics, but I knew infinitely less astronomy. I did not have her personal friendships with individual stars. She could describe fondly the eccentric behavior of certain spectral lines in a southern supergiant. (Much later she recommended to me the study of a peculiar young variable star, in the process of birth in a dense cloud of dust, because of its name – which aptly described its behavior – RU Lupi.) But her intuition as to what was important in astronomy was truly deep. By 1942 she had abandoned spectroscopic astrophysics and turned to the study of variable stars. She looked among them, also, for unifying principles, and for particular examples that illuminated the general trends; those topics are reflected in her autobiography briefly, at the end, with her marriage to Sergei Gaposchkin.

Returning to *The stars of high luminosity*, it can be seen as a more complex book, but with less mathematics and physics than *Stellar atmospheres*. It reflects her growing interest in the intrinsically brightest stars, which we now use extensively to obtain distances to other systems of stars, other galaxies. The same methods of atomic theory are applied, with more emphasis on the effect of pressure on the atoms, as well as temperature. Comparing a supergiant of the solar temperature with the Sun shows that the factor of 100 000 in intrinsic brightness results in subtle differences in the intensity and width of the spectral lines. The hydrogen lines are sharper in the supergiant, the ionized metals are stronger relative to the neutrals and certain elements are enhanced. Molecules are somewhat weaker. Most of these effects can be explained by lower pressure in the atmospheres; the supergiant is 300 times larger in radius, and although it may be 10 times more massive than the Sun, the force of gravity at the surface is only 0.0001 that in the Sun. Atomic collisions are less frequent, the atmosphere is more transparent and thicker, and there are second-order differences in the level of ionization of the metals. We have learned only recently the causes of the tendency to instability and to variability in light of these bloated stars, but they have an enormous practical advantage in being the brightest stars in our own Milky Way and in distant spiral nebulae. They can be recognized as individual stars in galaxies 30 million light years distant; groups of hot supergiants immersed in ionized gas clouds can be recognized at a fivefold greater distance.

The very high luminosity, complex evolutionary states and instability guarantee that supergiants will differ more from each other than do non-evolved stars. In her second book, Cecilia discusses typical examples and properties of several dozens of types of high-luminosity stars, and variables. In particular, two problems emerge. Some giants and supergiants have abnormal compositions. She mentions what are now known to be carbon-rich stars as a subclass of the variables; she explains upsilon Sagittarii, which has enormously strengthened helium lines, as having very high luminosity. Now we realize that carbon and helium are ashes of nuclear burning of hydrogen, and signs of genuinely abnormal composition. She wrote at too early a time to include nuclear physics and energy generation and transmutation in her thinking. Another problem, the apparent red color of distant hot stars (spectral types O and B), she faced with obvious discomfort. Her spectroscopic analysis had conclusively showed them to be hot, but their observed colors disagreed. Unfortunately, the proof that dust in space could absorb and redden starlight was published by R. J. Trumpler (1930) in the same year as her book. She refers only once to his paper. My own thesis (1937) was on the source and nature of the interstellar reddening of the stars of high luminosity. The idea that dust might redden distant stars was already in the wind, however; it was contradicted only by poor observations. Perhaps the most valuable lesson for us now is to believe only in good data, and to note the occurrence of strange phenomena, as Nature gives them to us, without undue concern for lack of explanation.

It is an odd sensation to review two books published 57 and 52 years ago; few except Shakespeare and The Bible can suffer such belated scrutiny. I wrote a brief biographical memoir about Cecilia for the American Philosophical Society, to which she was elected in 1936. That election was in itself remarkable – a woman, only 36 years old; these two major books appeared before she was 30. With 50 years of astrophysics having intervened, her work still reads well as serious science. It was fully conscious of the new world of atomic physics. It led forward to important problems in the study of nuclear astrophysics, as well as in the use of variable stars of high luminosity, probing the structure and rotation of our Milky Way, and the distances to other galaxies. Most important, it showed the bravery and adventure of a mind exploring the unknown with the available scientific apparatus and a complete belief in the power of human reason and logic.

Pasadena
26th October 1982

An historical introduction to 'The dyer's hand'

PEGGY A. KIDWELL
National Museum of American History, Smithsonian Institution

Cecilia Payne-Gaposchkin has been called 'probably the most eminent woman astronomer of all time'.[1] In 1923, after completing undergraduate studies at Newnham College of Cambridge University, she left England to do research at the Harvard College Observatory in Cambridge, Massachusetts. Within 2 years, Radcliffe College awarded her the first PhD in astronomy granted to a student at the Harvard Observatory. Her thesis, published as the monograph *Stellar atmospheres*, was widely acclaimed. She stayed on at Harvard, becoming Phillips Astronomer in 1938 and Professor of Astronomy and Chairman of the Department of Astronomy in 1956. Although she retired in 1965, she continued to do research until shortly before her death in 1979.

Toward the end of her life, Payne-Gaposchkin wrote the autobiography that follows. Astronomers, historians of science and those interested in women's role in academe will be intrigued by her story. Here she traces her lineage, describes her education, and outlines the beginning of her astronomical studies at Cambridge and at Harvard. She describes her later travels and research more briefly, closing with comments on diverse aspects of her career. I shall examine these events from a somewhat different perspective, that provided by Payne-Gaposchkin's published books and papers, and by correspondence from and about her. To keep this discussion within reasonable bounds, I will emphasize the years 1923–1925, the time Payne-Gaposchkin spent as a graduate student at the Harvard College Observatory.

The path to Harvard

Payne planned to study science from her youth, but was not at all sure precisely which science. At Cambridge, she settled on botany, chemistry and physics, with the greatest emphasis on physics. Then, at the end of her first year, she heard A. S. Eddington lecture on his eclipse expedition to the island of Principe and the implications of his results for the theory of relativity. She resolved to become an astronomer.[2]

At that time, most astronomy courses at Cambridge were taught as part of mathematics. Payne was officially a student in natural sciences and it was too late to change her course of study. She continued her work in other sciences but attended W. M. Smart's course on practical astronomy and Eddington's lectures on various theoretical topics. Under Smart's direction, she wrote a paper on the proper motion of stars in the cluster M36, and published her results in the *Monthly Notices of the Royal Astronomical Society*.[3]

At Cambridge, Payne also met the young astrophysicist E. A. Milne, and became a good friend of L. J. Comrie. Comrie, a native of New Zealand and a veteran of World War I, helped Payne put the telescope at Newnham College in working order, and gave her a book by Caroline Furness on variable stars to guide her first observations. Comrie also pointed out that women had more opportunities for astronomical research in the United States than in England. In May, 1922, he escorted Payne to the centennial meeting of the Royal Astronomical Society, where they heard a lecture by Harlow Shapley of the Harvard College Observatory. Much impressed, Payne decided to go to work at Harvard.[4]

Payne's willingness to leave her family and her native country to go to Harvard offered mute testimony to her passionate desire to do research in astronomy. Those who recommended her to Shapley described this eagerness. Eddington wrote

She has attained a wide knowledge of physical science including astronomy, and in addition possesses the valuable qualities of energy and enthusiasm in her work.[5]

Comrie worded his praise more strongly. After mentioning that he had thought it proper to nominate Payne to the Royal Astronomical Society, he wrote

> I know of no lady in England who is more likely to be successful at Harvard than Miss Payne. There is this to be said about her (between ourselves) – I believe she is the type of person who, given the opportunity, would devote her whole life to astronomy and that she would not want to run away after a few years training to get married.[6]

Payne would indeed devote her life to astronomy, although she did not believe that this precluded marriage.

Impressed by Payne's record and recommendations, Shapley offered her a small stipend.[7] Combining this with a fellowship from Newnham College, she had enough money to spend a year in the United States. She set sail in September, 1923. Comrie, who was teaching at Swarthmore College, met her in New York and helped her on the way to Boston; she soon was settled in Cambridge, Massachusetts.[8]

Payne found the atmosphere in Cambridge 'intoxicating'. She relished the New England climate, she enjoyed long talks with astronomers like Shapley and W. J. Luyten and, most especially, she treasured her freedom. As she later recalled, it was a great pleasure to do astronomy just as much as she wished, without having lectures to attend or classwork to worry about. Moreover, she could study as late as she pleased, with no rules about when the lights went out. She later said that 'for a bit, I almost worked night and day without stopping, it was marvellous'.[9]

Early research

Writing to Payne in March of 1923, Shapley had suggested that her greatest opportunities for rewarding research at Harvard lay in the photographic study of variable stars.[10] Payne replied that she was much interested in variable stars,[11] and Shapley recommended reading on the subject.[12] Once Payne arrived in Cambridge, however, Shapley rapidly discovered that she was even more interested in another topic, the physical interpretation of stellar spectra. During the late nineteenth and early twentieth centuries, Williamina Fleming, Antonia Maury, and Annie J. Cannon of the Harvard College

Observatory had examined photographs of the spectra of hundreds of thousands of stars. Noting patterns in the strength of absorption and emission lines, they were able to divide stellar spectra into seven broad classes, one class grading into the next.[13] This empirical classification did not depend on any theory about the physical constitution of the stars, and observers at different observatories disagreed about precisely what spectral characteristics should be used to classify certain stars. Alluding to a character in Lewis Carroll's *Through the looking glass*, the astrophysicist Henry Norris Russell described the situation in 1923 as follows

> At present, the ... classification meets only one of the White Knight's three specifications. We have decided what the *names* of the various classes shall be. We have next to define just what *shall be called* by each of these names. Before we do that, I think it will pay us to find out all we can about what they *are*.[14]

Payne and Russell's belief that it might be possible to account for differences in stellar spectra grew out of the work of the Indian astrophysicist M. N. Saha. In papers of 1920 and 1921, Saha argued that temperature and pressure determined the degree of ionization of atoms in stellar atmospheres.[15] Atoms of different degrees of ionization absorbed light of different wavelengths, hence the degree of ionization determined the spectrum of a star. Saha discussed in detail the conditions that would produce the faintest discernible or marginal appearance of an absorption line, as the spectral class at marginal appearance could be determined from published observations. Combining estimates of the pressure of stellar atmospheres with observations of marginal appearances, Saha arrived at a temperature scale for stars of different spectral class that agreed quite well with other proposals.[16]

E. A. Milne and R. H. Fowler at Cambridge[17] and Russell at Princeton[18] greeted Saha's work enthusiastically. They scrutinized his assumptions and considered the maximum intensities of lines as well as marginal appearances. Shortly before Payne left Cambridge, she learned that the subject of the next Adams Prize essay at Cambridge University was to be 'The physical state of matter at high temperatures'. The adjudicators of the Adams Prize suggested specifically that the essay might take the form of a contribution to the general theory of equilibrium of ionization, or it might aim at obtaining quantitative results with the aid of known physical data. Stellar spectra and other astronomical evidence could be discussed in this connection.[19]

The Adams Prize was open to anyone who had received a degree from Cambridge University.[20] In 1923, women were not yet granted degrees at Cambridge, so Payne could not compete officially. Nonetheless, she resolved to use available physical data and measurements of stellar spectra to test the theory of ionization, particularly as it applied to stellar atmospheres at high temperature.[21]

Payne's enthusiasm greatly impressed Shapley. A little over a month after her arrival, he confided in a letter to Russell 'she seems to be thoroughly capable of taking hold of the spectrum analysis as a department of the Observatory's work'. However, Shapley had already agreed to Russell's suggestion that Donald H. Menzel, a Princeton graduate student, should work over the Harvard spectra from the standpoint of ionization theory. Shapley now proposed that Menzel might concentrate on the lines of a few elements, particularly metals. He hoped to have 'some definite plan made early in the game, before anybody is imposed upon'.[22] In reply, Russell congratulated Shapley on getting Payne, and encouraged Menzel to concentrate on the lines of neutral metals. As these lines are observed primarily in stars of low temperature, this would leave the stars of high temperature for Payne.[23] Shapley concurred, but did little to prevent either Payne or Menzel from feeling imposed upon.[24]

Payne reports that once she began her measurements 'there followed months, almost a year as I remember it, of utter bewilderment'.[25] Despite these early struggles, Payne completed one paper within 6 months of her arrival at Harvard and three more during her first year there. By the time she received her doctorate in 1925, she had finished a total of six papers as well as a book-length monograph that drew together her results.

Initially, Payne avoided the metallic lines assigned to Menzel, although she did not confine herself to stars of high temperature. Her first paper presented measurements of the relative intensity of the lines of silicon in four states of ionization, tracing changes in intensity with spectral class. This gave a rough temperature scale for all but the very hottest stars.[26] By June, Payne had also measured changes in the intensity of carbon and helium lines in the hotter stars. These data gave a stellar temperature scale 'harmonious' with her results from silicon lines.[27]

Russell had pointed out that once the temperatures of stellar atmospheres were known, ionization theory would allow one to estimate the ionization potentials of atomic states not yet fully studied in the laboratory.[28] Following Russell's procedure, Payne estimated the ionization potentials of singly ionized oxygen, nitrogen and sulphur, and doubly ionized nitrogen, sulphur and carbon.[29] Menzel made similar calculations for neutral silicon and singly ionized scandium, titanium and iron.[30] Mindful of the interest of physicists in ionization potentials, Payne also prepared a synopsis of available data on ionization potentials. Here she included graphs of the change in ionization potential across rows of the periodic table and discussed what these patterns suggested about the arrangement of electrons in atoms.[31] In both her approach and her conclusions, Payne followed a recent paper of the Harvard physicist F. A. Saunders.[32] She did not include Menzel's results on this list.

Beginning in the spring of 1924, Payne seriously considered taking a formal degree at Radcliffe. Finding a degree program was a bit difficult, as neither Harvard nor Radcliffe had previously granted doctorates in astronomy. Shapley made the necessary arrangements, however, and Payne's written preliminary examinations were scheduled for June 10th, 1924. Harlan T. Stetson of the Massachusetts Institute of Technology was to provide questions in general astronomy, F. A. Saunders in physics, and Shapley in astrophysics.[33] Unfortunately, Stetson entirely forgot the exam; Shapley had to supply questions in astronomy as well as astrophysics. Writing to Saunders, he described the test as follows: 'I thought your questions were both fair and good. Those in astrophysics and astronomy were neither, but they were highly difficult and should suffice for the information we need.'[34] Of course Payne passed. Her reaction to the whole procedure is not recorded.

Payne may have found working day and night exhilarating, but her mother was concerned. In March, 1924, Mrs Payne wrote to Shapley asking that he be sure her daughter received a proper summer vacation. In a comment echoed by others who knew Payne later, her mother wrote

She is a healthy, but not really strong person, and lives largely on her enthusiasms, and while I delight to think of her doing the work she loves, I cannot help being anxious at times lest she should not allow herself the proper rest.[35]

Annie J. Cannon sent reassurances to Mrs Payne and Miss Payne spent most of August away from Harvard.

During her summer travels, Payne first attended the American Astronomical Society meetings in Hanover, New Hampshire. She was not only elected to the A.A.S., but gave a short paper on the astrophysical determination of ionization potentials.[36] She then went to the Toronto meeting of the British Association for the Advancement of Science. In Toronto, Payne took full advantage of the chance to see her old professors and to make new acquaintances. Russell and Saunders helped her meet such distinguished spectroscopists as J. S. Plaskett and A. Fowler, as well as the theoretician G. Lemaitre. She also talked to Milne. Although Payne had taken the first steps toward her doctorate and hoped to stay at Harvard, she wasn't sure how she would pay for further studies. Thus she listened with some interest when Robert G. Aitken asked her if she would like to spend a year at the Lick Observatory.[37] Payne also asked Eddington about the possibility of work in England. He said that the outlook for women astronomers was no better than it had been a year earlier and that the chance to study another year at Harvard 'was not to be lightly put aside'.[38] From Toronto, Payne went to Cleveland, Ohio for a complete rest. She wrote Shapley that she was 'having an excellent time here – very largely sleep', and even extended her stay by a week[39].

Back at the observatory, Payne discovered that she had been awarded the Rose Sidgwick Fellowship of the American Association of University Women. She wrote Aitken to decline the Lick Fellowship and plunged into work again.[40] A year earlier, Shapley had complained to Russell that Menzel's thesis plans seemed to incorporate all of spectrum analysis as related to temperature, the structure of matter, spectral series and ionization theory.[41] He may not have realized that Payne was equally ambitious. In any event, Menzel had finished his degree and Payne now set out to study the full range of stellar spectra. Menzel had estimated relative line intensities in the spectra of 20 stars of low temperature; Payne considered some 100 stars of this sort.[42] She found, as she had with silicon, that temperature increased steadily from the reddest stars toward the bluer ones. Ionization theory indeed might explain much about stellar spectra.

Results were most puzzling for precisely the stars that most interested Payne, those of highest temperature. Stars of this class, known as the O stars, characteristically show the lines of ionized helium in their spectra. In a paper of November, 1924, Payne presented her measurements of the absorption lines of 38 O stars. Like the Canadian astrophysicists H. H. Plaskett[43] and J. S. Plaskett,[44] she found that the Harvard classification of these stars did not correspond to gradations in temperature. H. H. Plaskett had suggested that the O stars be divided into several subclasses according to the relative intensity of their hydrogen and helium lines. In her paper, Payne argued against adopting Plaskett's subclasses at that time.[45]

As later paragraphs will indicate, Payne's forthright comments in 1924 probably did little to widen her opportunities for employment. However, careful study of H. H. Plaskett's paper may well have led her to the fundamental question of her thesis: how could ionization theory be used to determine the relative abundance of elements in stellar atmospheres? Some years earlier, Russell had compared H. A. Rowland's data on the relative intensity of absorption lines in the solar spectrum with F. W. Clarke's summary of the chemical composition of the Earth's crust. Russell found that the most common metallic elements of the solar atmosphere were virtually the same as the common metals of the Earth's crust. Moreover, there was a general similarity in the relative abundance of these metals. Russell concluded that if the Earth's crust were heated to the temperature of the Sun's atmosphere, it would give a very similar absorption spectrum.[46]

At the close of his paper on O stars, H. H. Plaskett sought to test Russell's views using observations of marginal appearances. Under questionable assumptions about the intensity of lines at marginal appearance and about the relative abundance of lithium and calcium in the Earth and the stars, Plaskett concluded that the relative abundance of barium, sodium, magnesium and hydrogen was quite similar in the Earth's crust and in stellar atmospheres. The atmospheres of stars contained a far greater proportion of helium than did the Earth's crust however.[47]

In December, 1924, Payne sent Russell the first draft of her own paper on the relative abundance of the elements in stellar atmospheres. Like Plaskett, she assumed that stars of all classes were chemically alike, differing in the degree of ionization of their atoms. However, she used Fowler and Milne's theory both to describe marginal appearances and to calculate relative abundances. Payne's paper shows the meticulous thoroughness that marks her work. Plaskett had used scattered observations of the marginal appearances of several elements. Payne used a relatively homogeneous collection of spectra, all taken on Harvard telescopes and examined by a single observer.[48] In her calculations, she used the stellar temperature scale she had deduced from ionization theory. Moreover, she obtained data on 20 of the 25 most abundant elements in the Earth's crust.[49]

Payne's results were startling. To be sure, atoms of silicon, carbon and common metals were in about the same relative numbers in stellar atmospheres and on Earth.[50] However, just as Plaskett had thought, helium was more abundant in the stars than on Earth. Moreover, hydrogen atoms, which in the Earth's crust were about three times more abundant than atoms of a common metal like aluminium, were apparently one million times more abundant in stellar atmospheres. In short, Payne's results suggested that stellar atmospheres were composed primarily of hydrogen and helium, with traces of other elements.

Shapley sent Payne's manuscript off to Russell, who at first glance thought that it seemed 'a very good thing'.[51] On further reflection, Russell suggested that Payne might reasonably raise her estimates of the abundance of magnesium and of iron, and noted data which would allow her to include potassium in her tables. Russell then commented to Payne

There remains one very much more serious discrepancy – namely that for hydrogen, helium and oxygen. Here I am convinced that there is something seriously wrong with the present theory. It is clearly impossible that hydrogen should be a million times more abundant than the metals.[52]

Russell went on to suggest that a theory of 'metastable' states, which he had developed with Karl Compton, might help to explain the apparent high abundance of hydrogen, although he did not know how this would apply to helium.[53]

Neither the original draft of Payne's paper nor any account of her reaction to Russell's letter has survived. However, early in February, when she sent the article to the *Proceedings of the National Academy of Sciences*, she commented that the stellar abundance deduced for hydrogen and for helium 'is improbably high, and is almost certainly not real'.[54] She said, just as Russell had suggested, that Russell and Compton's work on metastability might explain the great apparent abundance of hydrogen, and that helium remained a problem. Thus, although Payne's data indicated that the composition of stars differed radically from that of the Earth, she accepted Russell's authority, and argued that the stars were fundamentally like the Earth's crust.

Stellar atmospheres

Payne drew together her studies of stellar temperature, ionization potentials, spectral classification, and the relative abundance of the elements in a thesis entitled *Stellar atmospheres: a contribution to the observational study of high temperature in the reversing layers of stars*. This monograph, published for the Harvard College Observatory in the summer of 1925, established Payne's reputation as an astronomer. Russell wrote Shapley that it was the best doctoral thesis he had ever read, with the possible exception of Shapley's own thesis.[55] Thirty-seven years later Otto Struve and Velta Zebergs wrote in their history of twentieth-century astronomy: 'It is undoubtedly the most brilliant Ph.D. thesis ever written in astronomy.'[56] On a more mundane level, Shapley wasn't sure that as many as 100 astronomers and physicists would spend $2.50 for a book by a young astrophysicist.[57] He was delighted to report later that the entire edition of 600 copies sold out within three years.[58]

Stellar atmospheres was, as E. A. Milne put it, 'at the same time an attractive story and a work of reference'.[59] Payne divided her book into three parts. The first section was a lucid, systematic exposition of atomic physics as it pertained to the origin of line spectra, and of the physics of stellar atmospheres, independent of Saha's theory.

Here Payne supplied both necessary background for those unfamiliar with stellar spectroscopy and a valuable summary of recent data for experts in the field. She first outlined Bohr's theory of the atom, presented an updated table of ionization potentials, and gave a summary of recent data on the duration of excited states. Payne then listed physical conditions that influenced the appearance of absorption lines. The most important factors were temperature and pressure; hence she examined in some detail recent estimates of stellar temperatures and pressures. Payne next described the general features of stellar spectra, explaining several words commonly used. Finally, she discussed in turn earlier observations of the lines of each of those elements and compounds that had been observed or might be expected to appear in stellar spectra.

In the second part of *Stellar atmospheres*, Payne presented her physical interpretation of stellar spectra. After a general description of the excitation and ionization of atoms of a hot gas, she outlined the assumptions and relevant mathematical consequences of Fowler and Milne's version of Saha's ionization theory.[60] She also discussed evidence for the theory available from laboratory experiments of A. S. King and from studies of the solar spectrum by Russell and by Payne herself.

According to Fowler and Milne's theory, the temperature at which an absorption line has maximum intensity depends on pressure and on the excitation and ionization potential of the atomic state that produces the line. For each line she considered, Payne knew the requisite potentials and assumed that the pressure in stellar atmospheres was of the order of 10^{-4} atmospheres. Hence she could calculate the temperature at which each absorption line would be of maximum intensity. Observing the spectral class at which the line reached maximum, she deduced the temperature of the atmospheres of stars of that class. In her thesis, Payne presented her own observations of the relative intensity of lines produced by neutral atoms of 13 elements and by ionized states of atoms of 11 elements. Noting the spectral class at which each of these lines reached maximum intensity and calculating the temperature at maximum, she obtained a temperature scale for stellar atmospheres.

Menzel had found that for cooler stars, temperatures calculated from ionization theory were generally much higher than those computed by other methods.[61] To explain this observation, Fowler and Milne suggested that lines in the spectrum of a star were produced at different pressures. When most of the atoms of any element could be excited to produce a given line, the line would originate relatively high in the stellar atmosphere, where pressures were low. If, on the other hand, only a small fraction of the atoms of an element were in the state required to produce a line, that line would arise lower in the atmosphere, at a higher pressure. For example, Saha's theory indicated that in the Sun neutral calcium would be far more abundant than ionized calcium. Hence, Fowler and Milne argued, in calculating the temperature of the Sun's atmosphere, one should assume that lines of neutral calcium are produced at much lower pressures than lines of ionized calcium.[62]

Both the pressure at which a line reached maximum intensity and the fractional concentration of atoms in the state that produced the line varied with the temperature. Payne had originally assumed that pressures were constant for all lines in all stars and calculated temperatures. She now used spectrophotometric estimates of stellar temperatures to compute the pressures and concentrations found at the maxima of lines. In this way, she arrived at a consistent temperature scale for stellar atmospheres. As Russell commented to Shapley

To show that so much of the apparent discrepancy in the position of the maximum intensity is traceable to the relative abundance of atoms in the higher and lower excitation states, and to have the computed pressures come out as they do after this correction, is very nice indeed.[63]

After a brief discussion of the influence of absolute magnitude on stellar spectra, Payne turned to the final and shortest section of her monograph, 'Additional deductions from ionization theory'. Under this general heading, she included physical constants deduced astrophysically, as in her work on ionization potentials. She also indicated how ionization theory might create new problems in stellar classification, as with the O stars. Then Payne turned to the subject of the relative abundance of elements in stellar atmospheres, presenting the data and calculations from her 1925 paper. As before, she dismissed her results for hydrogen and helium as 'almost certainly not real'.[64]

Russell and Compton's work might possibly explain the observed abundance of hydrogen, but helium remained a mystery.[65]

In the final chapter of her thesis, Payne proposed several future problems of stellar spectroscopy – a revision of Rowland's tables of the solar spectrum, further analysis of the lines of nitrogen and other elements whose spectra were not yet understood, and an extension of observations to include infrared as well as visible light. The revision of Rowland's tables was a major undertaking of the Mt Wilson Observatory, completed in 1928.[66] Numerous physicists and astrophysicists have contributed to the better understanding of atomic energy levels. Payne herself hoped to study the infrared region of stellar spectra, but the problem was left to others.[67]

More generally, Payne affirmed her belief that in astrophysics 'observation must make the way for theory, and only if it does can the science have its greatest productivity'.[68] Her thesis amply demonstrated how empirical classification and rough measurement might pave the way to an understanding of stellar atmospheres. The subject of her research would change with the years, but observation, particularly as it pertained to theory, retained its central place.

Recognition and disappointment

During the years after she received her doctorate, Payne received many of the rewards accorded to promising young scientists. Her thesis was hailed as 'an excellent book',[69] 'a timely treatise',[70] and a 'book worthy of a place in every physical, as in every astronomical, library'.[71] In 1926, Payne became the youngest astronomer ever starred in J. M. Cattell's *American men of science*.[72] She also acquired an international reputation; in 1928 she was elected to the International Astronomical Union's Committee on Spectra.[73]

Several reviewers of *Stellar atmospheres* stressed that absolute measurements of the intensity of spectral lines were needed to extend Payne's observations of relative intensities.[74] The problem of spectrophotometric measurement of individual lines intrigued Shapley, and Payne won a National Research Fellowship to work on the topic.[75] She worked with Frank S. Hogg, the first male doctoral student at the Harvard College Observatory, and with an assistant provided by Radcliffe College.[76]

During these years, astrophysicists gradually came to believe that hydrogen was far more abundant in the stars than on Earth, just as Payne's calculations had indicated.[77] In 1925, Svein Rosseland dismissed Russell and Compton's ideas about the metastability of excited states of hydrogen, and argued on theoretical grounds that the element was repelled from the core of stars and concentrated in stellar atmospheres.[78] The following year, Russell and Charlotte E. Moore argued from observations of the width of lines in the solar spectrum that hydrogen was more abundant in the solar atmosphere than in the Earth's crust; they cited Rosseland's explanation of this phenomenon.[79] In 1928, Albrecht Unsöld argued that measurements of the contours of solar absorption lines could be used to estimate relative abundances. His calculation of the relative abundances of hydrogen and calcium agreed quite well with the figures Payne gave in *Stellar atmospheres*. To explain the enormous amount of hydrogen in the solar atmosphere, he too referred to Rosseland.[80]

That same year, Russell, Adams and Moore published a calibration of Rowland's scale for the intensity of lines in the solar spectrum, indicating how the number of atoms responsible for a spectral line varied with the Rowland intensity and the wavelength of the line.[81] To fix the scale, Adams and Russell compared the intensity of several lines in the solar spectrum with that observed in the spectra of giant stars. Looking at the data for hydrogen, they concluded that either the atmospheres of giants contained enormous amounts of hydrogen or they were not in thermal equilibrium.[82]

Eddington soon pointed out that Russell and Adams had misunderstood the effects of deviations from thermal equilibrium. At the same time, Menzel's eclipse observations suggested that the mean molecular weight of substances in the lower part of the Sun's atmosphere was only about 2. Reviewing this evidence and the data he had accumulated with Moore and Adams, Russell finally concluded that the enormous abundance of hydrogen in stellar atmospheres was a real phenomenon. In a masterly paper of 1929, he presented his data, and compared his estimates of the relative abundance of elements in the solar atmosphere with those Payne had obtained for the stars. Both astronomers had found vast amounts of hydrogen; more generally Russell found 'a very gratifying agreement' in these results. He did not mention that Payne had dismissed her data on hydrogen as probably spurious, nor allude to his role in shaping this conclusion.[83] Bengt Strömgren soon argued that hydrogen was the most abundant element in the core of a star as well as its atmosphere.[84] This fact would play a central role in the theory of the origin of solar energy that emerged in the late 1930s.

Neither the honors Payne received nor the growing awareness of the importance of her research spared her from two major difficulties encountered by women astronomers of the 1920s and 1930s. First, she was ineligible for academic appointments at all-male and many coeducational colleges, the schools that generally had the best observatories. Even some women's colleges, such as Radcliffe, had no women on the faculty. Secondly, astronomical observatories usually were located in isolated places. Those in charge often thought it was too dangerous to have women observe alone and improper for them to spend the night in the company of men. Hence women were barred from staff appointments and from observing. Unsurprisingly, most women in astronomy worked as computers, reducing large quantities of data accumulated for projects such as Harvard's *Henry Draper Catalogue* of stellar spectra.[85]

A few examples suggest how this discrimination affected Payne. In the fall of 1924, J. S. Plaskett wrote to Russell asking if he would recommend someone for a position at the Dominion Astrophysical Observatory in British Columbia, Canada. Plaskett preferred British or Canadian applicants, but was happy to consider others.[86] In reply, Russell noted that 'quite the best of the young folks' in astrophysics was Payne, who would complete her PhD the following spring.[87] Plaskett responded that 'there would be difficulty about the observing end of it with a woman in this isolated place and I think we can hardly consider her'. Plaskett also expressed disappointment in Payne's papers on O stars, but assured Russell that this was not the deciding factor in his rejection.[88]

Payne also encountered overt discrimination at Harvard. President A. L. Lowell did not believe that women should receive appointments from the Harvard Corporation.[89] Thus, when Payne taught part of the introductory graduate course in astronomy, she was not paid directly as an instructor or professor, but indirectly as Shapley's 'technical assistant'.[90] The courses she taught were not listed in the Harvard catalogue until 1945.[91] Moreover, Payne had some difficulty observing on even Harvard-owned telescopes, particularly at the Boyden station in South Africa. As Shapley explained to one correspondent, the director of the station feared that he would not get along well with Payne, and also thought that 'a lone woman would be in danger from the blacks'.[92]

Payne was by no means oblivious to these restrictions. In 1930, when Shapley urged her to abandon stellar spectroscopy and to take up E. S. King's work on photographic photometry, she was especially distressed. She complied, but when Virginia Gildersleeve approached her about leaving Harvard to teach physics at Barnard College, she took the possibility seriously.[93] In an unusually frank letter to Russell, Payne outlined her problems as follows

> ... I have during the past four years had a very unhappy time at Harvard; the chief reasons have been (a) personal difficulties within the Observatory particularly with Dr Shapley, and usually arising out of personal jealousies because he seemed to like others more than myself. (b) **disappointment because I received absolutely no recognition, either official or private, from Harvard University or Radcliffe College; I cannot appear in the catalogues;** I do give lectures, but they are not announced in the catalogue, and I am paid for (I believe) as 'equipment'; certainly I have no official position such as instructor. Presumably this is impossible, and so I have always thought it; but I have felt the disappointment nevertheless. (c) I do not seem to myself to be paid very much; quite honestly I think I am worth more than 2300 dollars to the Observatory. (d) In the seven years I have spent at Harvard I have not got to know any University person through my work (which confines my acquaintance to the Observatory staff and Professor Saunders); whereas the wife of any Harvard man of my status is called upon by the wives of dozens of others.[94]

Payne thought that she had already overcome her first problem, and hoped a Barnard offer would allow her to barter for a better position at Harvard. Specifically, she wanted three things

> an increase of salary; three months 'vacation' in the summer (during which I should be *free to work*, preferably at some other Observatory, just as Plaskett does); and some attempt to assist me into the company of my

'intellectual equals'. The third demand is unspecific and unsatisfactory, but it represents a genuine need.[95]

Payne asked Russell whether he thought her demands were reasonable and likely to be granted, as she didn't want to act unwisely.

Russell's papers contain no reply to this letter. Payne was interviewed for the Barnard job, but some doubted that she was meant to teach undergraduates and she apparently was never offered the position.[96] Nevertheless, the threat served its purpose. Shapley had suggested in 1929 that Payne should not be considered for a professorship at Columbia because she 'could not well afford scientifically to detach herself from a large observatory'.[97] He now took several steps to assure that she appreciated this affiliation. Record books of the Harvard College Observatory indicate that she received an increase in salary to $2700 per year. This still left her pay between that of starting computers, who received about $1000 a year, and new lecturers on the astronomy faculty, who earned about $3000 annually.[98] To negotiate any raise at all in 1931 was no small achievement however.

Shapley also tried to provide Payne with more time for research and to improve her status at Harvard. He hired Gladys Wickson to edit Observatory publications, freeing Payne to spend some summers away from Harvard during the 1930s.[99] She did not regularly do research at any other observatory however. As to Payne's hope for greater recognition in the Cambridge community, Shapley nominated her to the Harvard Faculty Club, and she was elected an associate member in the fall of 1931.[100] Shapley also proposed that Payne's course on variable stars be included in the Harvard catalogue, listing his name as instructor for official purposes. President Lowell and the Dean of the Faculty rejected this idea outright.[101] Lowell resigned in 1933 and his successor, J. B. Conant, proved more amenable to the idea of Corporation appointments for women. Annie J. Cannon and Payne-Gaposchkin both received such appointments in 1938. Payne-Gaposchkin also won recognition outside of Harvard, with election to the American Philosophical Society in 1936 and, at Russell's urging, nomination to the National Academy of Sciences in 1939.[102] More immediately, the deaths of Adelaide Ames and Bill Waterfield in the early 1930s, growing national and international tensions, and

Payne's marriage to Sergei Gaposchkin in 1934 may possibly have drawn her attention to other matters.[103]

In conclusion, I wish to emphasize four themes of Payne-Gaposchkin's early career that appeared again in her later work. First, she was passionately devoted to astronomy, willing to forego convention, native land and fortune to 'move among the stars'.[104] She was a talented musician, a scholar of both classical and modern literature, and an enthusiastic traveller. She would also be an occasional historian, a loyal wife and the mother of three children.[105] Most especially, however, she sought to unravel the mysteries of the stars.

Secondly, Payne-Gaposchkin had a thorough understanding and intimate knowledge of astronomical observations, particularly of stellar spectra and variable stars. Elske v. P. Smith has recalled 'her way of referring to individual stars, even spectral lines, as familiar and distinctive friends'; others have made similar comments.[106] In *Stellar atmospheres*, Payne-Gaposchkin was able to bring Saha's theory to bear on observations of stellar spectra; variable stars required a more empirical survey of the problem.

Thirdly, Payne-Gaposchkin had a talent for lucid exposition, manifest in scientific treatises like *Stellar atmospheres*, and in her correspondence. She later demonstrated this ability in her lectures and in books for a popular audience. To quote one former student

> I listened to a course of hers in variable stars, and what I learned was the English language. If there was ever anyone who speaks it in the way it was intended to be spoken, it was Cecilia and it was beautiful.[107]

Readers may judge this talent themselves in the book that follows.

Lastly, Payne-Gaposchkin's sex limited both the instruments available to her and her opportunities for professional advancement. By the early 1950s, her lowly status at Harvard was particularly incongruous. She taught courses, advised students and, with Sergei Gaposchkin, supervised a considerable number of assistants. She was also an officer of the Council that directed the Harvard College Observatory and served occasionally as acting chairman of the astronomy department. In 1956, she was finally made a full professor, the first woman at Harvard promoted to this rank. She sent handwritten notes to all the women students in astronomy, inviting them to celebrate the occasion in the library. After appropriate flowery speeches, Payne-Gaposchkin herself rose to speak. According to Nannilou Dieter Conklin's account

she said 'I find myself cast in the unlikely role of a thin wedge'. It brought down the house. She was a very large person, but she could make fun of herself and see the humor of the whole business.[108]

Payne-Gaposchkin was a large woman in many ways indeed, and we are lucky to have her story.

Notes

1 Owen Gingerich, Leo Goldberg, Fred I. Whipple and Charles A. Whitney, 'Cecilia Helena Payne-Gaposchkin: memorial minute' *Harvard University Gazette*, **76**, 1981.
2 See 'The dyer's hand', Chapter 7.
3 C. Payne, 'Proper motion of the stars in the neighborhood of M36 (NGC 1960)', *Monthly Notices of the Royal Astronomical Society*, **83**, 334, 1923.
4 In addition to Payne-Gaposchkin's account below, I have used an interview of Payne-Gaposchkin by Owen Gingerich that took place March 5, 1968. The interview is available in typescript at the Niels Bohr Library, American Institute of Physics, New York, NY.
5 The letter of recommendation accompanies C. Payne to H. Shapley, February 27, 1923; H. Shapley Directorial Papers, Harvard University Archives, Cambridge, MA. This and other letters from the Shapley Papers are cited by permission of the Harvard University Archives.
6 L. J. Comrie to H. Shapley, March 7, 1923; H. Shapley Directorial Papers, Harvard University Archives.
7 H. Shapley to C. Payne, April 16, 1923; H. Shapley Directorial Papers, Harvard University Archives. Shapley awarded Payne an Edward C. Pickering Fellowship. The Pickering Fellowships had been established at the Harvard College Observatory in 1916 by a gift of the Maria Mitchell Association, a group associated with Vassar College. The awards were designed to provide women working at the observatory with funds for their astronomical research. A few women had done graduate work there previously, notably Henrietta Leavitt (1892–93), Annie J. Cannon (1895–97) and Arville D. Walker (1914–17). None had obtained a degree. Astronomy 20, a graduate research course, was first listed in the Radcliffe catalogue in 1920. Adelaide Ames was the first student in 1922–23. In the fall of 1923, Payne and Harvia Hastings Wilson joined Ames in this fledgling graduate pro-

gram. Ames finished her A.M. in 1924, becoming the first woman to receive a graduate degree in astronomy at Radcliffe. On the early Pickering fellowships, see Lydia S. Hinchman to H. Shapley, January 22, 1929; H. Shapley Directorial Papers, Harvard University Archives. The names of early graduate students in astronomy at Radcliffe are included in the *Alumnae Directory of Radcliffe College*, Cambridge, MA: Radcliffe Alumnae Office, 1968. I wish to thank the Radcliffe Archives for information about enrolment in early graduate courses.

8 C. Payne to H. Shapley, June 23, 1923 and September 19, 1923; H. Shapley Directorial Papers, Harvard University Archives.

9 C. Payne-Gaposchkin, Interview with Owen Gingerich, March 5, 1968; Niels Bohr Library, American Institute of Physics.

10 H. Shapley to C. Payne, March 12, 1923; H. Shapley Directorial Papers, Harvard University Archives.

11 C. Payne to H. Shapley, April 5, 1923; H. Shapley Directorial Papers, Harvard University Archives.

12 H. Shapley to C. Payne, April 16, 1923; H. Shapley Directorial Papers, Harvard University Archives.

13 On the early history of spectral classification, see 'The dyer's hand', Chapter 10, as well as: S. I. Bailey, *The history and work of Harvard Observatory*, Harvard Monograph no. 4, New York: McGraw-Hill, esp. pp. 149–59, 1931; B. Z. Jones and L. G. Boyd, *The Harvard College Observatory: the first four directorships*, Cambridge: Harvard University Press, esp. pp. 218–44, 1971; and D. DeVorkin, 'A sense of community in astrophysics: adopting a system of spectral classification', *Isis*, 79, 29–49, 1981.

14 H. N. Russell to H. Shapley, October 30, 1923; H. N. Russell Papers, Manuscripts Division, Princeton University Library, Princeton, NJ. This quotation and others from letters in the Henry Norris Russell Papers are published with permission of Princeton University Library.

15 Earlier astronomers had suggested that temperature might determine the spectral class of a star. See Agnes Clerke, *A popular history of astronomy during the nineteenth century*, London: Adam and Charles Black, pp. 421–24, 1887; and D. DeVorkin, 'A sense of community in astrophysics: adopting a system of spectral classification', *Isis*, 79, 29–49 (esp. pp. 30–6, 43), 1981. For H. N. Russell's early views, see his paper 'Relations between the spectra and other characteristics of the stars', *Nature*, 93, 227–30, 252–8, 281–6, 1914.

16 M. N. Saha. 'Ionization in the solar chromosphere', *Philosophical Magazine*, 40, 472–88, 1920; 'Elements in the sun', *Philosophical Magazine*, 40, 809–24, 1920; 'On a physical theory of stellar spectra', *Proceedings of the Royal Society*, 99A, 135–53, 1921. D. V. DeVorkin and R. Kenat 'Quantum physics and the stars. 1: The establishment of a stellar temperature scale. 2: Henry Norris Russell and the abundance of the elements in the atmospheres of the sun and stars',

Journal of the History of Astronomy, **14**, 102–32, 180–222, 1983.
17 See E. A. Milne, 'Ionization in stellar atmospheres', *The Observatory*, **44**, 261–9, 1921; and E. A. Milne and R. H. Fowler, 'Intensities of the absorption lines in stellar spectra ...', *Monthly Notices of the Royal Astronomical Society*, **83**, 403–25, 1923.
18 H. N. Russell, 'The theory of ionization and the sun-spot spectrum', *Astrophysical Journal*, **55**, 119–44, 1922.
19 *The Observatory*, **46**, 127–8, 1923.
20 For a statement of the terms of the Adams Prize, see W. D. Niven (ed.), *Scientific Papers of James Clerk Maxwell*, vol. I, p. 288, Cambridge: Cambridge University Press, 1890.
21 See 'The dyer's hand', Chapters 11 and 12, and C. Payne-Gaposchkin, Interview with Owen Gingerich, March 5, 1968; Niels Bohr Library, American Institute of Physics. R. H. Fowler won the Adams Prize. His essay was the basis of his book *Statistical mechanics, the theory of the properties of matter in equilibrium*, Cambridge: Cambridge University Press, 1929.
22 H. Shapley to H. N. Russell, October 25, 1923; H. N. Russell Papers, Princeton University Library.
23 H. N. Russell to H. Shapley, October 30, 1923; H. N. Russell Papers, Princeton University Library.
24 H. Shapley to H. N. Russell, November 3, 1923; H. N. Russell Papers, Princeton University Library. For Payne's reaction to Menzel's arrival, see Chapter 12 below. For an account of Menzel's view, see D. H. Menzel, Autobiography; D. H. Menzel Personal Papers, Harvard University Archives. I wish to thank Mrs D. H. Menzel for allowing me to examine this autobiography.
25 See 'The dyer's hand', Chapter 12.
26 C. Payne, 'On the absorption lines of silicon in stellar atmospheres', *Harvard College Observatory Circular*, no. 252, 1924.
27 C. Payne, 'On ionization in the atmospheres of the hottest stars', *Harvard College Observatory Circular*, no. 256, 1924.
28 H. N. Russell, 'The theory of ionization and the sun-spot spectrum', *Astrophysical Journal*, **55**, 119–44 (esp. pp. 140–4), 1922.
29 C. Payne, 'On ionization in the atmospheres of the hottest stars', *Harvard College Observatory Circular*, no. 256, pp. 6–8, 1924.
30 D. H. Menzel, 'A study of line intensities in stellar spectra', *Harvard College Observatory Circular*, no. 258, p. 14, 1924.
31 C. Payne, 'A synopsis of the ionization potentials of the elements', *Washington National Academy Proceedings*, **10**, 322–8, 1924.
32 Compare Payne's paper with F. A. Saunders, 'Some aspects of modern spectroscopy', *Science*, **59**, 47–53, 1924.
33 H. Shapley to Theodore Lyman, May 20, 1924, and T. Lyman to H. Shapley, May 22, 1924; H. Shapley Directorial Papers, Harvard University Archives.

34 H. Shapley to F. A. Saunders, June 10, 1924; H. Shapley Directorial Papers, Harvard University Archives.
35 Mrs Edward Payne to H. Shapley, March 22, 1924; H. Shapley Directorial Papers, Harvard University Archives. For later indications that Payne sometimes worked on the verge of exhaustion, see H. Shapley to P. J. van Rhijn, June 17, 1933; Harlow Shapley Directorial Papers, Harvard University Archives; and N. Wiener to C. Wiener Franklin, August 27, 1925; Norbert Wiener Papers, Institute Archives, M.I.T., Cambridge, MA. Interviews with Frances Wright and Charlotte Moore Sitterly also suggested both the strength of Payne's enthusiasms and the strain that this sometimes put on her health.
36 For an abstract of Payne's paper, see *Publications of the American Astronomical Society*, **5**, 179–80, 1926.
37 C. Payne to R. G. Aitken, September 12, 1924; R. G. Aitken Papers, Lick Archives, University of California, Santa Cruz, CA.
38 C. Payne to H. Shapley, August 10, 1924; H. Shapley Directorial Papers, Harvard University Archives.
39 C. Payne to H. Shapley, August 23, 1924; H. Shapley Directorial Papers, Harvard University Archives.
40 C. Payne to R. G. Aitken, September 12, 1924; R. G. Aitken Papers, Lick Archives. For further information about the Rose Sidgwick Fellowship, see Margaret Malby (ed.), *History of the fellowships awarded by the A.A.U.W. 1888–1929*, New York: Columbia University Press, pp. 8 and 89, (1929?).
41 H. Shapley to H. N. Russell, October 25, 1923; H. N. Russell Papers, Princeton University Library.
42 D. H. Menzel, 'A study of line intensities in stellar spectra', *Harvard College Observatory Circular*, no. 258, 1929, and C. Payne, *Stellar atmospheres: a contribution to the observational study of high temperatures in the reversing layer of stars*, Harvard Monograph no. 1, Cambridge, England: W. Heffer & Sons, p. 119, 1925.
43 H. H. Plaskett, 'The spectra of three O-type stars', *Publications of the Dominion Astrophysical Observatory*, **1**, 325–86, 1922.
44 J. S. Plaskett, 'The O-type stars', *Publications of the Dominion Astrophysical Observatory*, **2**, 287–358, 1924.
45 C. Payne, 'On the spectra of class O stars', *Harvard College Observatory Circular*, no. 263, 1924.
46 H. N. Russell, 'Solar spectrum and the earth's crust', *Science*, **39**, 791–4, 1914. For a detailed study of changing ideas about relative abundances, with particular emphasis on Russell's ideas, see articles of David DeVorkin and Ralph Kenat in the *Journal for the History of Astronomy*.
47 H. H. Plaskett, 'The spectra of three O-type stars', *Publications of the Dominion Astrophysical Observatory*, **1**, 356–63, 372–81, 1922.

48 Payne's data were not entirely homogeneous, as she used spectra taken on two different telescopes at Harvard and on the 13-inch Boyden telescope in Arequipa. See C. Payne, *Stellar atmospheres*, p. 116. Payne's own observations are recorded in the Observing Book of the 24-inch Reflector, no. H8, Plate Stacks, Harvard College Observatory, Cambridge, MA. I thank Martha Liller for showing me these records.

49 C. Payne, 'Astrophysical data bearing on the relative abundance of the elements', *Washington National Academy Proceedings*, **11**, 192–8, 1925. Three of the five elements that Payne omitted had not been observed in the stars; the ionization potentials of the remaining two were not yet known.

50 The spectra of the two most abundant terrestrial elements, oxygen and nitrogen, were not yet sufficiently understood for Payne to calculate the relative abundances. She later did this in *Harvard College Observatory Bulletin*, no. 835, published in 1926.

51 H. N. Russell to H. Shapley, December 9, 1924; H. N. Russell Papers, Princeton University Library. I thank Owen Gingerich and David DeVorkin for drawing my attention to this letter in the Russell Papers, as well as others cited in the following notes. Gingerich and DeVorkin's own comments on the study of relative abundances are cited in notes 46 and 77.

52 H. N. Russell to C. Payne, January 14, 1925; H. N. Russell Papers, Princeton University Library.

53 See K. T. Compton and H. N. Russell, 'A probable explanation for the behaviour of hydrogen in giant stars', *Nature*, **114**, 86–7, 1924.

54 C. Payne, 'Astrophysical data bearing on the relative abundance of the elements', *Washington National Academy Proceedings*, **11**, p. 197, 1925.

55 Quoted in H. Shapley to C. Payne, August 15, 1925; H. Shapley Directorial Papers, Harvard University Archives.

56 Otto Struve and Velta Zebergs, *Astronomy in the 20th century*, New York: Macmillan, p. 220, 1962.

57 H. Shapley to H. N. Russell, August 6, 1925; H. N. Russell Papers, Princeton University Library. Up to that time, publications of the Harvard College Observatory had been distributed to other observatories free of charge.

58 H. Shapley to Ada Comstock, September 14, 1928; H. Shapley Directorial Papers, Harvard University Archives.

59 E. A. M[ilne], 'Stellar spectra and the physics of gases at high temperatures', *Nature*, **116**, 530, 1925. Other reviews of *Stellar atmospheres* are listed in notes 69, 70, 71 and 74.

60 Here Payne used R. H. Fowler and E. A. Milne, 'The intensities of absorption lines in stellar spectra', *Monthly Notices of the R.A.S.*,

403–24, 1923; and Fowler and Milne, 'The maxima of absorption lines in stellar spectra', *Monthly Notices of the Royal Astronomical Society*, **84**, 499–515, 1924.
61 D. H. Menzel, 'A study of line intensities in stellar spectra', *Harvard College Observatory Circular*, no. 258, esp. pp. 15–17, 1924.
62 R. H. Fowler and E. A. Milne, 'The maxima of absorption lines in stellar spectra', *Monthly Notices of the Royal Astronomical Society*, **84**, 499–515, 1924.
63 Quoted in H. Shapley to C. Payne, August 15, 1925; Harvard University Archives.
64 C. Payne, *Stellar atmospheres*, p. 188.
65 *Ibid.*, pp. 56–7, 153–89.
66 C. E. St. John et al., *Revision of Rowland's preliminary tables*, Carnegie Institution of Washington Publication, no. 396, 1928.
67 In *Stellar atmospheres*, Payne considered wavelengths of 3900 to 4500 Angstroms. Spectroscopists soon exceeded these limits. For example, the *Revision of Rowland's preliminary tables* value for the solar spectrum extended from 2975 to 10218 Å.
68 C. Payne, *Stellar atmospheres*, p. 200.
69 A. Kohlschütter, *Die Naturwissenschaften*, **14**, 139–40, 1926. Kohlschütter's phrase was 'ausgezeichneten Buches'.
70 P. W. Merrill, *Publications of the Astronomical Society of the Pacific*, **38**, 33–4, 1926.
71 J. Q. Stewart, *Physical Review*, **26**, 870, 1925.
72 S. S. Visher, *Scientists starred 1903–1943 in 'American men of science'*, Baltimore: Johns Hopkins Press, p. 377, 1947. The practice of starring the names of distinguished scientists was discontinued after 1943.
73 H. Shapley to A. Comstock, July 26, 1928; H. Shapley Papers, Harvard University Archives.
74 In addition to the reviews of P. W. Merrill and J. Q. Stewart mentioned in notes 70 and 71, see A. C. D. Crommelin, *Journal of the British Astronomical Association*, **38**, 305–6, 1926. Reviews which concentrated on other matters were O. Struve, *Astrophysical Journal*, **64**, 204–8, 1926; and *The Times Literary Supplement*, December 17, 1925, p. 876.
75 H. Shapley, 'Preliminary report on the brightness of absorption lines', *Harvard College Observatory Bulletin*, no. 805, 1924.
76 Payne and Hogg's papers on photometry are listed in the bibliography. The funds provided by Radcliffe College for Payne's assistant are noted in a memo in the Ada Comstock Papers at the Radcliffe Archives, Schlesinger Library, Radcliffe College, Cambridge, MA. The memo is filed under 'Payne' for the year 1932–3. See also H. Shapley to A. Comstock, September 14, 1928; H. Shapley Directorial Papers, Harvard University Archives.

77 For a summary of these developments, see K. R. Lang and O. Gingerich (eds), *A source book in astronomy and astrophysics, 1900–1975*, Cambridge: Harvard University Press, pp. 243–4, 1979. Gingerich has discussed this topic further in a talk given to the Orion Nebula Symposium in New York in December 1981. For a more detailed study, see the articles by DeVorkin and Kenat mentioned in note 46.

78 Svein Rosseland, 'On the distribution of hydrogen in a star', *Monthly Notices of the Royal Astronomical Society*, **85**, 541–6, 1925.

79 H. N. Russell and C. E. Moore, 'On the winged lines in the solar spectrum', *Astrophysical Journal*, **63**, 1–12, 1926.

80 A. Unsöld, 'Ueber der struktur der Fraunhoferschen Linien und die quantitative Spectralanalyse der Sonnenatmosphäre', *Zeitschrift für Physik*, **46**, 765–81, 1928.

81 H. N. Russell, W. S. Adams, and C. E. Moore, 'A calibration of Rowland's scale of intensities for solar lines', *Astrophysical Journal*, **68**, 1–8, 1928.

82 W. S. Adams and H. N. Russell, 'Preliminary results of a new method for the analysis of stellar spectra', *Astrophysical Journal*, **68**, 9–36, 1928.

83 H. N. Russell, 'On the composition of the sun's atmosphere', *Astrophysical Journal*, **70**, 11–82, 1929.

84 B. Strömgren, 'Opacity of stellar matter and the hydrogen content of the stars', *Zeitschrift für Astrophysik*, **4**, 118–52, 1932.

85 For a general account of the role of women in American science, see Margaret Rossiter, *Women scientists in America, struggles and strategies to 1940*, Johns Hopkins University Press, 1982. Rossiter has described the pattern of women's work in American science between 1880 and 1910 in '"Women's work" in science, 1880–1910', *Isis*, **71**, 381–98, 1980. Pamela Mack has examined the role of women in American astronomy from 1895 to 1920 in 'Straying from their orbits: women in astronomy in America', in *Women of science*, G. Kass-Simon and Patricia Farnes (eds.), Bloomington: Indiana University Press, pp. 72–116, 1990. For a general discussion of recent trends, see Debby J. Warner, 'Women's astronomers', *Natural History*, May 12–26, 1979.

86 J. S. Plaskett to H. N. Russell, October 27, 1924; H. N. Russell Papers, Princeton University Library.

87 H. N. Russell to J. S. Plaskett, November 4, 1924; H. N. Russell Papers, Princeton University Library.

88 J. S. Plaskett to H. N. Russell, January 18, 1925; H. N. Russell Papers, Princeton University Library. Plaskett mentioned Russell's recommendation to Shapley in a letter of November 14, 1924. In reply, Shapley proposed another candidate. See H. Shapley to J. S.

Plaskett, December 9, 1924. Both of these letters are in the H. Shapley Directorial Papers, Harvard University Archives.

89 Williamina Fleming had held the Corporation appointment of Curator of Astronomical Photographs from 1899 to 1911; however, President Lowell thought that this was an unfortunate precedent. See P. Mack, 'Straying from their orbits', pp. 94, 100–2.

90 H. Shapley to the Harvard University Bursar, May 4, 1929 and March 24, 1930; H. Shapley Directorial Papers, Harvard University Archives.

91 C. Payne-Gaposchkin to H. N. Russell, September 24, 1945; H. N. Russell Papers, Princeton University Library.

92 H. Shapley to Alice Farnsworth, April 26, 1940; Alice Farnsworth Papers, College History Collection, Williston Library, Mt. Holyoke College, South Hadley, MA. For the objections of the director of the Boyden Station, see a cablegram from J. S. Paraskevopoulos to H. Shapley, May 30, 1933; Correspondence concerning the Boyden Station, Harvard College Observatory Papers, Harvard University Archives.

93 V. C. Gildersleeve to C. Payne, December 8, 1930, and C. Payne to V. C. Gildersleeve, December 11, 1930; Dean's Office Correspondence, File 41 (Physics Department), Barnard College Archives, Barnard College, Columbia University, New York, NY.

94 C. Payne to H. N. Russell, December 11, 1930; H. N. Russell Papers, Princeton University Library.

95 Ibid.

96 V. C. Gildersleeve to C. Payne, December 13, 1930, and George G. Pegram to V. C. Gildersleeve, March 3, 1931; Dean's Office Correspondence, File 41 (Physics Department), Barnard College Archives.

97 H. Shapley to George G. Pegram, August 30, 1929; H. Shapley Directorial Papers, Harvard University Archives. Curiously, Shapley was quite willing to recommend his colleague H. H. Plaskett and Otto Struve of the Yerkes Observatory for the Columbia position, despite their needs for access to telescopes. I thank Karl Hufbauer for drawing this letter to my attention.

98 Payne's salary and those of the computers are listed in the Payroll of the Harvard College Observatory, Harvard University Archives. In 1927, H. H. Plaskett was asked to come to the Harvard College Observatory at a salary of $3000 per year. See H. Shapley to F. A. Saunders, September 21, 1927; H. Shapley Directorial Papers, Harvard University Archives. When D. H. Menzel came as an assistant professor in 1932, he received $4500 a year. See D. H. Menzel to H. N. Russell [spring, 1932], H. N. Russell Papers; Princeton University Library.

99 H. Shapley to H. H. Plaskett, June 18, 1931; H. Shapley Directorial Papers, Harvard University Archives. From 1939, poor health forced

Wickson to give up much of her editorial work and Payne-Gaposchkin took on increasing editorial responsibilities. See correspondence between H. Shapley and Gladys Wickson; H. Shapley Directorial Papers, Harvard University Archives. These letters date from 1939 to 1942.

100 H. Shapley to G. W. Cram, July 27, 1931 and G. W. Cram to H. Shapley, November 23, 1931; H. Shapley Directorial Papers, Harvard University Archives.

101 H. Shapley to C. H. Moore, February 7, 1931 and C. H. Moore to H. Shapley, February 13, 1931; H. Shapley Directorial Papers, Harvard University Archives.

102 H. N. Russell to H. Shapley, November 24, 1937; H. N. Russell Papers, Princeton University Library. See also H. Shapley to C. Payne-Gaposchkin, April 25, 1940; H. Shapley Directorial Papers, Harvard University Archives. Payne-Gaposchkin was never elected to the National Academy of Sciences.

103 See 'The dyer's hand', Chapter 17.

104 See 'The dyer's hand', Chapter 9.

105 For comments about Payne-Gaposchkin's diverse interests, I am particularly grateful to her daughter, Katherine Haramundanis, and to Frances W. Wright, Owen Gingerich, Charlotte Moore-Sitterly, Vera Rubin and Charles A. Whitney. On Payne-Gaposchkin's interest in modern literature, see Helen Heinemann, 'Cecilia Payne-Gaposchkin', *Radcliffe Quarterly*, **66**, 38, March, 1980. In addition to her autobiography, Payne-Gaposchkin's historical writings include obituaries of several astronomers and 'The Nashoba plan for removing the evils of slavery: letters of Frances and Camilla Wright, 1820–1829', *Harvard Library Bulletin*, **23**, 221–51, 429–61, 1975.

106 Elske V. P. Smith, 'Cecilia Payne-Gaposchkin', *Physics Today*, **33**, 65, June, 1980.

107 Nannilou Dieter Conklin, Interview by David DeVorkin. A typescript of this interview is available at the American Institute of Physics.

108 *Ibid.*

Acknowledgments

Here I wish to thank Payne-Gaposchkin's family, colleagues and acquaintances for sharing their correspondence and recollections with me. I also much appreciate the friendly assistance of a widely scattered group of archivists, particularly Jean Preston and Clark Elliott. Finally, I thank Karl Hufbauer, Vera Rubin, Owen Gingerich, Katherine Haramundanis and Jesse Greenstein for their comments on an earlier draft of the text. This research was made possible by a grant from the American Philosophical Society.

References

Hoffleit, Dorrit 1991. 'The evolution of the Henry Draper Memorial', *Vistas in Astronomy.* **34**, 107–62.

Hoffleit, Dorrit 1993. *Women in the history of variable star astronomy.* Cambridge, Mass.: American Association of Variable Star Observers.

Kidwell, P. A. 1984. 'Women Astronomers in Britain, 1780–1930.' *Isis*, **75**, 534–46.

Kidwell, P. A. 1986. 'E. C. Pickering, Lydia Hinchman, Harlow Shapley and the Beginning of Graduate Work at the Harvard College Observatory.' *The Astronomy Quarterly.* **5**, 157–71.

Kidwell, P. A. 1987. Cecilia Payne-Gaposchkin: Astronomy in the family. *Uneasy careers and intimate lives: women in science 1790–1979*, Pnina Abir-Am and Outram (ed.), New Brunswick, NJ: Rutgers Univ. Press, pp. 216–238.

Kidwell, P. A. 1992. Harvard astronomers and World War II – disruption and opportunity. *Science at Harvard University: historical perspectives*, Clark A. Elliott and Margaret W. Rossiter (eds.), Bethlehem: Lehigh University Press.

Lankford, John and Rickey L. Slavings. March 1990. Gender and science: women in American astronomy, 1859–1940. *Physics Today*, **43** No. 3, 58–65.

Mack, Pamela E. 1990. 'Straying from their orbits: women in astronomy in America', *Women of Science*, G. Kass-Simon and Patricia Farnes (eds.), Bloomington: Indiana University Press; pp. 72–116.

A personal recollection

KATHERINE HARAMUNDANIS

Personality

No printed page, no book, can ever adequately distil the essence of any personality so well and so aptly that the reader sees more than the merest shadow of that personality; but the meagre shade, when formulated and constructed with care, may perhaps preserve something of the loved individual.

A family member is undoubtedly the worst possible person to write a biographical essay, as she brings a lifetime of recollections and prejudices with her to the writing. However, I feel the importance of this story is so great that I need give no apologies for its writing.

This is the story of a woman – who happened to be my mother – and who by the work of her own hand was unique. She was truly a Renaissance woman, witty, scholarly, patient, a paragon of scientific virtue, an indefatigable research enthusiast, a domestic wife. Erudite in several languages, a world traveler, a book lover, she was also an inspired cook, a marvelous seamstress, an inventive knitter and a voracious reader. Though her days were filled with lectures and scientific work, her evenings revolved around the kitchen, where she turned out countless tantalizing meals, her needlework, experiments with woodworking, bookbinding, cardplaying, puzzles, or simple entertaining. Bedtime was always accompanied by 'something to read'.

In appearance she was tall, with a dignified bearing. She wore her hair short, rarely set, and she almost never wore makeup. With her height (about 5′ 10″ in her forties), fine proportions, and slight plumpness, her appearance was regal and classical; some would compare her favorably to the Venus de Milo. She had regular, even beautiful features, a slim nose, grey-blue eyes, thin lips, a wide brow, and high cheekbones. In her later years, she generally wore glasses (for far-sightedness and astigmatism). She dressed conservatively, sometimes mannishly, and had little interest in clothing for its own sake. Her preferred color, however, was orange, and some of her favorite dresses were bright with this hue. Although she was a handsome woman, she cared little for feminine niceties. She never followed the foibles of fashion, nor engaged in time-consuming attempts to retain youth.

Frugal but not parsimonious, she saved money for things she wanted passionately, like a trip to London, by skimping on clothes and household things. It seemed more meaningful for her to spend money on an intangible like a trip, than on a permanent object like a table or a refrigerator.

Although she never appeared to have a systematic means of filing things, she always knew which pile contained the item she needed. She had tremendous nostalgia for family heirlooms, and she suffered agonies if a china cup or dish was broken. Mended with care, the chipped and cracked crockery remained in daily use or sat on a shelf to be admired.

She was fluent in German and French, although she always felt her French pronunciation was poor, and she read Italian, Russian, Latin and ancient Greek as the whim struck her. She learned a language to use it, to read something that could not be read in translation rather than as an exercise. Icelandic was a minor challenge, though I cannot say that she truly mastered it. Dante, in the original Italian, was one of her favorites.

She was addicted to puns, made them in several languages, and generally met those made by others with a suppressed chuckle. Her subtle wit was much a part of her charm. It was a rarity to hear her laugh out loud. I never saw her cry. These traits, so important in her character, seemed indelibly part of her, but later I realized that they probably accrued from her English upper-class upbringing and a stern mother.

Keenly interested in cooking, she was happiest trying out a recipe she had not used before and, such was her expertise, she rarely had a failure. She was able to make the transition from French to British to oriental cooking without trouble, often insisting that the only thing preventing her from making a particular dish was lack of the proper ingredients. Such was her enthusiasm for cooking that after every dinner she prepared the kitchen was a shambles.

Spices were a great favorite with her and she accumulated a wide variety. With her spices, she combined her expert knowledge of botany with her practical side. She often noted that with spices we consume every part of the plant: seeds from dill and celery, leaves from bay and oregano, stems from parsley, bark from cinnamon, flowers from saffron. Cloves, allspice, star anise, mustard, all held a place in the crowded and fragrant spice shelf over her kitchen counter. An expert in using her spices, she never over- or under-flavored a dish.

She was uniformly warm towards all of us children, although she did not relate easily to the very young. She was pleased at our later, intellectual successes and grieved deeply at our unhappinesses. If we were in a period of difficulty, she would refer to herself as akin to the grieving Demeter, Greek goddess of agriculture, who let all the world lie dormant while her kidnapped daughter Persephone was sequestered in Hades.

She had a tendency to feel that she could influence external events in our favor; but we also were taught that we were responsible for our own actions. In discussing human frailties she might remark 'anyone can make a mistake, but only a fool makes the same mistake twice'.

We gave her the greatest happiness when conversing with her on an intellectually challenging subject such as astronomy, mythology, literature or the classics. Her own preference was to avoid discussion of emotionally charged or sensitive issues, perhaps saying 'let's talk about it later', a 'later' that never came.

She left the stronger forms of discipline to our father, who believed in spanking, but her method of discipline with a stern look and a set countenance was as effective as physical means. No child of hers could doubt when she was displeased. But she could temper a scolding with her warm wit: when we were too rambunctious she would say, smacking her palm on the table, 'I want silence and very little of that'.

She was supportive but uninvolved in our school achievements. We knew that she expected us to do well because she expected us to be able to converse intelligently on a subject we had studied. Thus, in an indirect manner, we understood one of the aims of learning: to be able to communicate one's learning to others.

As a wife, she seemed able to communicate with my father, although they had some conflicts. By common agreement he controlled the funds, paid the bills and did much of the housework. She did the cooking, sometimes the washing up, some of the marketing and the washing, darning, and sewing. Sewing or knitting tasks she sometimes took on for her own amusement, perhaps working up a new pattern or creating a new design. For example, she designed and created a heavy pullover on which she knitted the first few bars of *Jingle bells*.

At home, as a family, we often played bridge together but I was mystified by the bidding and always secretly relieved to be declared 'dummy', for then I could sneak off to a book until the hand was played out. I believe my parents made a sound bridge pair and enjoyed playing bridge together as partners.

Many years later, my mother introduced me and my husband John to the delights of 'Scat', a three-handed card game with an ingeniously complex system of bidding, and exotic terms such as 'Matador' and 'Ramsch'. I found it much more fun than bridge.

Mother enjoyed solitaire immensely and had several favorites. Perhaps the one she played most often was 'the demon', which she later expanded to play with two decks. This she dubbed the 'giant squid'. In her later years, playing a hand of the 'giant squid' with my father at her side to kibbitz, was her last task of the day. This was a good wind-me-down to relax her before bedtime.

In my parents' household, frugality reigned. The army surplus store was a well-exploited source of quality goods. In a burst of feminine enthusiasm, my mother purchased a surplus parachute; its yards and yards of uncontrollable gossamer white nylon, then an expensive rarity, were turned into practical underthings. Its miles of heavy nylon cord did service for years.

My mother rarely used the telephone and had a great dislike for the instrument. Even when you could get her on the telephone, she was uncomfortable and in haste to get off. No matter how modern or sophisticated the equipment, she did not care for it. She extended this dislike for modern contrivances to her laundry activities. It was many years before my parents obtained an automatic clothes washer; they never did own a dryer.

In discussion, she was eminently reasonable. You could discuss almost any subject with her and never be reduced to verbal fisticuffs. This was one of her very finest attributes. She never left anyone in a conversation feeling slighted, even when she disagreed with him, and she let one feel entitled to one's opinion; she was never unreasonable. I always felt she had no prejudices, but in later years I thought I detected some.

She did not encourage us to study any special subject. What we studied, what our career goals were, was left to ourselves. She might comment on the practicality of a decision, she might question a course of study if it seemed vague or without goal, but she never categorically stated that a particular course of study was pointless. She once remarked to me, in a discussion of careers, that if she had not become a scientist she would probably have become an actress, clearly not a Hollywood-style actress, but an English stage actress.

Her interest in the theatre seemed generated by those early visits to the Old Vic and other London theatres; her interest in science seemed encouraged by her scientifically inclined relations, particularly the Horners and Dora Pertz, the Edwardian interest in natural history, and her adored botany teacher Miss Dalglish; her love of travel was inspired by early journeys.

When she set herself a task, she was indefatigable. Her powers of concentration were so great that she could work for hours without stopping. This, moreover, was not only on occasion. She could sustain this level of activity for long intervals, days or even weeks at a time, until the problem had been solved or the chaos reduced to order. Occasionally miffed because someone commented on the piles of papers with which her office was strewn as she buried herself in a problem, a set of light curves to analyze, or a paper to write, she would mutter crossly, 'My office may be disorderly but my mind is not!', and of course she was quite right.

Her one vice was cigarettes, which she smoked for perhaps 50 years, with rare attempts to give them up. They were, in the end, to take their toll. Once she arrived in her office for the day, particularly when working headlong as was her wont, she chain-smoked, lighting one cigarette from the last. Ashes were strewn broadcast, absent-mindedly, and the ashtray overflowed.

The war years

My younger brother Peter, born in 1940, had been a source of anxiety to my parents, for he did not talk. Their anxiety led to many difficulties and there was unhappiness all around. Visits to doctors of all kinds were to no avail. The doctors could not agree on a diagnosis. Deafness, fear, mental retardation; there was no concensus. Finally, late in 1943, Peter startled us all, coming downstairs in the morning singing 'From the halls of Montezuma'. Of course then we all knew it was over – this was a Great Event and naturally followed by Great Rejoicing, but scars were carried for many years from those early anxieties. Peter, of course, went on in life to write a PhD thesis on the Friedman Universe, so one can conclude that he came through it all with flying colors.

My mother was excruciatingly busy during the war years, lecturing, writing papers, travelling to meetings, with or without us. My parents also bought a farm, with rocky, piney soil, in northern Massachusetts, which they dubbed the Highland Farm. It was fitted out as a chicken farm, holding at its height over 3000 chickens, 200 turkeys (in season), two pigs, a cow, a sheep, and a gaunt and beautifully ugly workhorse. We all worked hard on the farm, although as a child I could not see all the problems it must have entailed. I believe their original idea for the farm was to provide a living for a refugee family, but they could never find a family that seemed to understand how to deal with so many chickens.

The farm was a good experience for us children; we did farm work and lived in primitive conditions in a one-room cottage my father constructed in the pine woods. We had a tiny wood stove to cook on, used an outhouse, and hauled water half a mile from the farmhouse. In a burst of Russian enthusiasm, my parents dubbed the little house in the woods the 'Izbushka', which I understood meant 'little hut on chicken legs' straight out of the Russian folk tale about Baba Yaga.

In spite of these conditions and the bloodthirsty mosquitoes and deer-flies of the hot summer, my mother was undaunted. In the earliest years she cooked on the wood stove, the 'Pet', and produced not only delicious meals for the ravenous crew, but even managed to bake bread in the tiny, drafty oven. This seems like a remarkable achievement, although at the time, in the ignorance of youth, I rather took it for granted. After electricity was brought to the cabin, she could cook on a hot plate, which made summertime cooking much more reasonable.

When my maternal grandmother died in 1947, we received a great shipment of things from her London home at Horbury Crescent. Particularly important was a large quantity of books that came from the family libraries of the Paynes, the Wilkinsons and the Pertz's. As both Mr Payne (my mother's father) and Mr Pertz (her maternal great-grandfather) had been historians, much of the library contained historical works. There was also a great deal of literature: French, German, Latin, Greek, a little English, and a sprinkling of other languages, such as Italian, Spanish and Icelandic. These books seemed to open the doors of civilization.

There were also a few pieces of furniture, some china, tableware, and paintings, mostly by my grandmother. One of my favorites is her copy of Turner's 'Chichester Canal', a mellow, sunset painting with a masted schooner in still water reflecting the glowing sky and the faint spire of Chichester Cathedral in the distance.

My mother travelled a great deal, even during the hectic war years. She gave countless lectures to colleagues and to interested laymen. She went annually to the American Philosophical Society. She attended meetings of the Astronomical Society and went periodically to Yerkes Observatory in Wisconsin. She sang in the Lexington Choral Society and taught in the Sunday School of the Unitarian Church, carrying on a girlhood tradition.

I recall her characteristic determination in the face of adversity. One winter Sunday, it was so bitterly cold that the old, black Ford would simply not start. But car or not, she had to give her Sunday School class. She pulled on a pair of thick woollen slacks, bundled up warmly, and set out on foot to walk the $3\frac{1}{2}$ miles to Lexington Centre. I can still see her walking down Shade Street, snowdrifts piled high on either side, in her dark coat and hat. She passed from view on turning the corner. She was an example of determination and commitment.

In 1942 my parents traveled to Mexico to attend the opening of the Tonanzintla Observatory. They returned with lovely souvenirs, serapes and Mexican majolica. I believe the trip was a Great Event but we were glad to have them back.

During the war, my parents had to contend with gasoline and food rationing, and blackouts. In an attempt to conserve gasoline, they sometimes bicycled to Cambridge along Massachusetts Avenue. I do not recall that they did this often, however.

As part of the war effort, my parents followed the President's directive to grow vegetables and provide food for the country. Apparently they both took this task very seriously and made a cooperative effort of it. Part of their contribution (over 50000 eggs and numerous broilers) came from the Highland Farm. In Lexington they engaged in another cooperative effort, growing and canning vegetables. My father prepared the garden, later turned back into a bowling green, and my mother canned or pickled the produce.

One Thanksgiving (1941) from which a menu survives, there was a 14 lb turkey from the farm, potatoes, beans and cabbage from the garden, a mince pie of local ingredients (baked by myself), and a pumpkin pie from a pumpkin grown by Edward. I believe Dr Prager, a German astronomer, then a refugee, and Dr Whipple were guests at this dinner.

My mother was meticulous in her canning technique, perusing all the informative government bulletins, examining each bottle carefully, boiling everything. Days were spent standing in the steaming kitchen, peeling fruit, preparing vegetables, washing bottles, scalding jar rings; the results were jar after jar of delectable food. Albert the sheep, of late tethered in the back yard happily cropping grass in a neat circle, ended in jars, and the two little black-and-white piglets at the farm turned into bacon and spare ribs.

Those ribs provided succulent, surreptitious snacks for us children. My mother packed them with sauerkraut, in good German style, in casks stored in the cellar. We would sneak downstairs, open the wooden cask, and sneak out a couple of ribs. Of course there was Big Trouble when our nefarious theft was discovered and the cask was found empty of all ribs.

My parents were characteristically generous when offering shelter to a Japanese-American family who had been interned. Reverend Horikoshi, his wife and their children Eliot and Nancy, came to live with us. As the house on Shade Street was already full to bursting with the five of us, a live-in maid and a boarder in the attic, the Horikoshis settled themselves in the cellar. At least the cellar was warm, and Mrs Horikoshi was exceptionally neat, but one can hardly imagine that it was easy for them. However, even those minimal accommodations must have seemed better than a concentration camp.

When my brother Edward entered the first grade, I was so keenly disappointed at being too young to attend that my mother, to assuage my unhappiness, took me on a trip to the island of Nantucket off Cape Cod. I was entranced by the ferry ride to the island and enjoyed our walks over the cobbled streets and along the beach. There was even a charming lighthouse and my cup overflowed when we went to dinner, at my extravagant request, in a fine restaurant situated on a boat ('The Skipper') tied up at the wharf. It seemed like such a very Great Adventure though in truth Nantucket is not very far from Boston.

The Observatory

The Harvard Observatory seemed very much the centre of my mother's life. Both my parents were there six days a week, sometimes spent the evening there if there was a lecture or an 'Open House' for the public, and engaged in ancillary activities there. My mother started the Observatory Philharmonic Orchestra (OPO) and conducted frequent rehearsals and its first concert. The orchestra was small but enthusiastic; it played classical pieces, sometimes abridged, such as Mozart's 'Jupiter' symphony. After my mother relinquished her association with the orchestra, Frances Wright kept it going for many years. My parents' foray into political action through their Forum for International Problems was also carried out in the lecture hall of the Observatory.

At the Observatory, for us children, there were a large garden to play in, a steep hill for sledding in winter, the plate stacks with a spookily delightful spiral staircase, and a creaky dumbwaiter. There was also the dome of the 15-inch telescope, frigid in winter, with a chair on a tall scaffold that went squeakily up and down as the seated rider turned a large wheel in his lap. Beneath the circular rotunda of the telescope lay its great stone footings surrounded by a dusty catacomb, perfect for hide-and-seek.

My mother's office, on the top floor of the Brick Building (Building D), was tiny and almost inaccessible for to reach it one had to run the gauntlet of all the open office doors of the scientists, including the forbidding inner sanctum of the director's office. My father's office was more accessible as we could reach it by two routes, either through the warren of assistant's cubbyholes, or through a ground-floor window. Mother's office was perennially crammed with papers, books and a typewriter. My father was usually found with an eyepiece in his hand, or to his eye, with an astronomical photographic plate on the light lecturn before him. Perhaps an assistant would be seated across from him, recording his estimates of star brightness as he examined the plates.

At Christmas time, Dr Shapley gave grand parties for the entire Observatory. There were delicious delicacies served by uniformed helpers, a large, beautifully decorated Christmas tree, and communal Christmas carols. Dr Bok had a marvelous melodious baritone voice and sang *Good King Wenceslas*, singing the part of the Good King, while Miss Chubb, who had a fine, if shrill, soprano voice, sang the part of the Page. But these family-like parties and the elegant residence where they took place, came to an end a few years after the war.

At the Observatory, there were both local astronomers and visitors. Old Mr Campbell had an office reached through a tiny museum. Its glass cases were filled with beautiful sundials and astrolabes, and large celestial globes set on pedestals stood about. He seemed grim and forbidding, with his great shock of white hair and quiet ways; of course, he was not really forbidding, but only seemed so.

Dr Shapley had a rather craggy face and an extraordinary, intense personality. His lovely wife Martha, the gracious hostess of his many parties, seemed a permanent institution. Dr Bok, with his deep voice and strong accent, seemed a resplendent and exotic character. His tales of the murderous thunderstorms in South Africa were frightening. Dr Menzel seemed always handsome and urbane, while his beautiful wife Florence was perennially cheerful, friendly and high-spirited. Dr Russell, a tall, elderly gentleman, was so remote that his principal attraction for me was his large grandfather clock with a moon dial, that stood in the foyer of his home at Princeton. It seemed a fitting clock for a Great Astronomer to possess.

Frances Wright, a long-time friend of my mother, had a great gift for liking young people. Dr Whipple, elegant and polished, and his sympathetic and ebullient wife Babby were well liked by both my parents. As Director of the Smithsonian Astrophysical Observatory during the early, hectic years of the space program, Dr Whipple's inspired leadership brought the tiny, obscure institution into the limelight where it made lasting contributions to astronomy and geophysics.

Dick Thomas, dashing and excitable, gave the impression of being constantly in motion. His wife Gladys seemed perennially cheerful and understanding. They were often at the Observatory in the early years. Henrietta Swope was a slim woman with long hair tied in braids around her head. During the war years, she was a frequent visitor to our house in Lexington.

Dorrit Hoffleit was an astronomer for whom my mother had the very greatest respect. No one, mother said, understood spectra as well as Dorrit. While Dorrit did not often visit the Observatory, she was a visitor in my parents' home when she did. Her beautiful long, brown hair, bound up at the nape of her neck, was a trademark.

Charlotte Sitterly, a charming, scholarly woman, and her quiet husband Banny, were early visitors in Lexington. She and my mother had a long lasting friendship and shared a mutual admiration and respect.

George Gamov visited memorably, a jovial, blustery man with a strong accent and a marvelous sense of fun. He spent an hour with us children making simple polyhedra (pyramids, cubes, dodecahedra) out of folded paper. He was a scientist who managed also to be a remarkable human being.

The Schwarzschilds sometimes came on a visit. Martin was perenially handsome and elegant with an infectious equanimity; his wife Barbara had equal charm. He was another remarkable scientist who remained a remarkable human being.

Jesse Greenstein and his lovely, vivacious wife Naomi seemed infrequent but memorable friends. He too, kept the touch of humanity in his scientific pursuits.

Among non-scientists, the beautiful and disciplined Doris Beaudoin was memorable. She first came to the Observatory out of high school and proved so competent and efficient that she eventually became the Director's secretary. A devout Catholic, she had a remarkable and celibate existence. Her equanimity and unremitting service were only to be admired. Miss Harvey, an elderly, portly lady who befriended me in the plate stacks, was extraordinarily kind and left the impression of a sympathetic lady of infinite patience and no ambition.

My mother had intense feelings concerning her colleagues. While astronomy was her life, astronomers were its lifeblood. She admired, even revered, certain luminaries such as Shapley, Struve and W. W. Morgan. She had a high opinion of many, some of whom were her students, for example, Dick Thomas and Anne Underhill. The greatest criticism she could make of anyone was that his or her work was 'thrown together' or 'slapdash'; her greatest praise was to say of someone's work that it was 'really fine'.

Occasionally she succumbed to an immoderate burst of jealousy when a colleague trespassed on a piece of 'her' subject, but she was genuinely delighted when a real advance was made.

She recorded in an early diary[1] when she had temporarily given up all hope of studying science, that the headmistress of her school, a specialist in classical languages, had encouraged her to take examinations in Ancient Greek with the objective of competing for the Gilchrist medal. This suggestion filled her with happiness and anticipation, in strong contrast to the despair she had felt earlier that same school term at being unable to pursue scientific studies. This hope for the recognition a prize of stature would bring her was an underlying theme of her existence. Unfortunately, the jealousy she sometimes felt towards her colleagues did not go unnoticed.

Nevertheless, astronomers admired her uniformly, some extremely so. Struve praised her lavishly when he described her thesis[2] and when he reviewed her book *The galactic novae*.[3] Pannakoek mentioned her contribution to the quantification of spectral analysis;[4] Lang and Gingerich utilized excerpts from her doctoral thesis in their

A personal recollection

compendium,[5] where her work stands with classic contributions by Saha, Russell, Einstein and others. Öpik described her career and talents in glowing terms

> The meteoric brilliance of the life track of this genius of an English girl, many-sided, excelling in pioneering astronomical research as well as in music...[6]

Hoffleit recorded that she was

> unquestionably the greatest woman astronomer of the mid-twentieth century. Hers represents the acme of scientific writing that is both good science and excellent literary composition.[7]

Whitney called her 'an astronomer's astronomer',[8] and Mihalas praised both her scientific and personal attributes

> we shall all notice the loss of her leadership, wisdom, kindness, and warmth. She was truly one of the great astronomers of this century, and an absolutely extraordinary human being.[9]

Morgan, a luminary in his own right among the lights of the astronomical community, had the highest praise for her

> For a long time I have considered Cecilia Payne-Gaposchkin to be one of the truly great astronomers of the Twentieth Century. This conclusion could have been reached on the basis of her Doctoral Thesis 'Stellar Atmospheres' alone; but I have been in a particularly good situation to recognize and appreciate the greatness of the woman herself – even under the pall of the negativism of Harlow Shapley.[10]

Elske v.P. Smith wrote of her

> She may have felt that she had walked with giants; now we recognize that she, herself, was one.[11]

Travels

My mother had a tremendous love for travel. In some ways, many of her happiest moments seemed to occur on a journey. This interest and enthusiasm seemed inspired by early travels. For example, entranced by a visit to the Netherlands when she was 11, she preserved a record of the journey in a short essay.[12] Short excerpts show her kindled enthusiasm

Our first holiday on the Continent! Imagine Charing Cross, on one of the last days of July, a few minutes before the 9 o'clock train started. All was bustle and excitement . . .

Later after the channel crossing to Ostende

Then the boat entered the harbor, and moved SO slowly, that we seemed to be stationary, with the station and pier moving slowly back. Oh! here was the Continent indeed! . . .

The family stayed for the summer in Belgium, with excursions to Bruges, dikes and beaches. On their return she recorded

The sun shone brightly on the vanishing shores, farewell! farewell! to the Continent, till – ? . . .

This enthusiasm for seeing new places, viewing new wonders, new museums, new people, remained with her all her life; the powers of observation and description seen in this early essay were never to be lost. Her words also evince the charm that was so much a part of her character.

Later youthful travels during World War I were confined to England. She recorded in a lone diary holidays in Charmouth and Buck's Mills on the coast. After school was out in July, with many goodbyes to school friends, she spent three days packing and then travelled to the holiday site.

(1 August 1916, Tuesday)
Came to Buck's Mills. 10-hour journey. By train to Bideford and then by carriage – i.e., high waggonnette to Buck's Mills – 9 miles. A slow drive with a most talkative Devonshire farmer. The most beautiful place I have ever seen. In a dear little valley, wooded, with the sea down below.

(2 Wednesday)
Morning, painted flowers. Afternoon went for a walk on the Downs. Beautiful wild country. Great sea-mists.

(6 Sunday)
Went to Church. Alas! What singing. H. (Humfry) was convulsed.[13]

In every journey she took she seemed always to preserve some special place, some important incident. These memories were as much a part of her character as her scientific pursuits.

In 1921 and 1922 she traveled to Italy, Austria and Germany with her family. Although the Paynes were far from wealthy, Mrs Payne somehow managed to provide for her children all the important amenities of an upper-class household. Dilys Powell, Humfry's widow, paid Mrs Payne a high compliment in the brief biography she wrote about Humfry: 'She brought up the boy and his two sisters by a miracle of courage and self-sacrifice.'[14] Mrs Payne had a strong influence on all her children, both by insistence on excellence, living up to their heritage as 'being a Payne', and determination in the face of privation and adversity. This heritage was an indelible part of my mother's character.

A journey of lasting consequence in my mother's life, her long pilgrimage in 1933 to Pulkova with the return leg through Hildesheim and Göttingen, she has described in 'The dyer's hand'.[15] A diary of this trip (not in her possession when she composed her autobiography) recorded her visits to several observatories and sights that she saw. For example, with the Lundmarks she took a train for Landscrona *en route* to the site of Tycho Brahe's observatory

(Saturday, 8 July 1933)
An hour crossing the flat and fertile countryside of Skåne. . . . Down to quay, and take boat for Hven. Tiny boat, only link with mainland.

After the short journey, with Elsinore on the horizon

Hot and dusty walk along country road. Farmhouses, thatched and timbered, round three sides of a court, in whose center is a well. . . . At last we come to a railed enclosure, site of Tycho Brahe's observatory. Nothing of buildings left – were excavated and covered in again . . . On to site of Uraniborg. Few remains here. Two alchemist's furnaces, a well, and a private dungeon. It seems that when Tycho Brahe left the house was destroyed by the islanders who rightfully hated him . . .[16]

She recorded her perplexity with unfamiliar languages (Swedish, Finnish, Russian), troubles with her expired visa to the Soviet Union (solved by the American Consul in Tallinn as she was travelling on a US passport), and the grim conditions at the Pulkova Observatory. The astronomers worked under conditions of both physical and intellectual hardship, remote from most of the astronomical community, and were immensely pleased at her visit. She had conferences with many, including the 78-year-old Belopolsky, Tikhoff and

I. Lehmann (Mrs Balanowsky)
who is a very remarkable woman and a fine scientist. Tried to inspire her to further work on Novae (her spectra, taken by Tikhoff, are SUPERB).

Her reception in Germany was more formal, but she was clearly relieved to be back on familiar territory.

(Saturday August 5, Göttingen)
Attended scientific sessions all the morning, and met Gaposchkin, who turns out to be the dominant fact of the A.G. [Astronomisches Gesellschaft] meeting so far as I am concerned. His history and present tragic situation have no place in this diary of mine, but it does seem as if it is my place to find some way out for him, as he can no longer work in Germany. At the request or desire of Lindblad, Lundmark, Kopff, and Guthnick, and with the support of Eddington, I must do everything that is possible.[17]

Her efforts were to be rewarded and several months later she received the briefest note from her aloof and much-admired Eddington

Dear Miss Payne:
I am very glad to hear of your rescue of Gaposchkin.
Agathe Schwarzschild is settled at Newnham and evidently delighted with Cambridge. She was in to tea on Sunday, and seemed very pleased with the way things had turned out. So that also has been a successful intervention.[18]

Her travels did not always have such a lasting effect on her life, but her love of travel never waned.

In 1948 we five all went on a great trip to the West Coast. My parents were to work for the summer at Cal Tech and they took us on a marvelous tour of the country in the process. Each family member chose a National Park or a scenic wonder to visit and a route was planned that touched them all. To keep expenses down, mother prepared our lunches at roadside parks and we camped or stopped for the night in modest cabins. After the trip was over we computed the cost per mile for transportation by car; it was 2 cents a mile.

The trip west touched the Great Smoky Mountains, Mammoth Cave, Grand Canyon, Zion Canyon, Boulder Dam and Lake Mead. In New Mexico we stopped at White Sands Proving Ground near Alamagordo to see the firing of a V2 rocket. While in California we went to the opening of Mount Palomar, and I recall vividly a spectacular view of Saturn seen through the 100-inch reflector. The colorful planet seemed near enough to touch. On our return east we visited Lick Observatory, Yosemite Valley, Crater Lake in Oregon, and the Cascades. Wherever we stopped I practiced my specialty, Falling Into Streams. My mother always scolded me gently when I came back from yet another expedition Wet Again, but she never reminded me the next time to be careful. This was a marvelous quality in a mother.

From Oregon we drove north to visit the observatory at Victoria, B.C., reaching it with a lovely late summer ferry-ride through Puget Sound; later we turned east through the Snake River Valley in Idaho, one of the loveliest valleys in the world. At Yellowstone we saw all the usual wonders, the geysers and hot springs, and I was startled by a large moose standing in front of the little museum. I thought she was a statue and marveled at how lifelike she was!

The following year my mother took me to Europe. We traveled across the Atlantic on the SS *Batory* of the Polish–American line, stopping first in England. Shipboard life in the five-day crossing seemed frivolous and boring, although the food was extraordinary and the seamen were courteous. As we had a terrific storm at sea, mother spent most of the time in her bunk, but I found a friend, Ellen, and we wiled away the time playing cards in the empty lounge.

Once in England, I put a terrific crimp in my mother's plans to attend a Garden Party at Buckingham Palace by coming down with pneumonia. After I was better, we visited my mother's Aunt Florence, who was very thin and very old. She looked as though she lived on two lettuce leaves a day. She lived alone in a tall townhouse and had the peculiar habit of locking all the doors of the house from the inside, with the idea that, should a burglar come in through a window, he would at least not be able to get into the rest of the house. Mother later informed me that she was really 'quite mad'.

The Astronomer Royal, Sir Harold Spencer Jones, invited us for a week to Herstmonceux Castle where I was entranced to be living in a fifteenth-century castle with crenellated battlements and staircases in the turrets. From there mother took me to the Roman Wall at Housesteads, the lovely village of Reeth, and Newcastle-upon-Tyne where we attended an extremely crowded sit-down dinner party. Later we traveled to Bath to visit my mother's sister Leonora and her husband Walter. They had a lovely home filled with Wedgwood escutcheons in the woodwork. Leonora and Walter were both architects and took us about to see some of the marvelous architectural edifices of that elegant city.

We then took the channel steamer to France and stopped in Paris where my mother attended a conference at the Observatory. We stayed in a grand hotel and were feted at Versailles in an elegant champagne reception. In Paris we also had lunch with Jeff Myers, whose daughter was later to marry my older brother Edward. Leaving Paris, we took a train over the St Gothard pass to Italy. Italy seemed like a second home to my mother, although it could not have held the attraction England did for her.

A stop in Pisa followed our exit from the long overnight ride to Milano. In Pisa we visited the Leaning Tower, the bell tower of the Cathedral, which I religiously climbed, and the Cathedral itself. In the Cathedral hung a large bronze lamp on a long chain attached high above to the arched ceiling. My mother pointed out to me that it was the swaying of this lamp that gave Galileo the idea of the pendulum.

From Pisa we entrained for Florence where we stayed for several days. We visited the Cathedral where she showed me the marvelous Ghiberti doors on the baptistery, the Pitti Palace, elegant residence of Luca Pitti, crushed by the Medicis, and the Loggia dei Lanzi, an open-air portico which housed Cellini's bronze statue of Perseus with the head of Medusa. Perseus grasps the snake-like hair, holding the head at arm's length. My mother pointed out that Cellini had taken as much care with the back of Medusa's head as with her face, even though it was normally unseen by the spectator; this was the sign of a true craftsman. As the statue stood under an overhanging roof, I waited long hours in the nearby Piazza until the sun was low enough to shine on Perseus' face, to take a much wanted photograph.

Also in Florence, we visited the Uffizi Gallery, where my mother could indulge her passion for Duccio and Cimabue, early painters of religious subjects, and I could begin to appreciate the beauty of Botticelli. There, too, we saw the magnificent tomb of the Medicis, with the marvelous, muscular sculptures of Michaelangelo, and

Michaelangelo's David, standing solitary at one end of a tiny room, its nearest companions the unfinished sculptures of contorted captives. This was my earliest recollected exposure to truly great sculpture and it left a lasting impression.

Mother also took me to the little museum that housed the collection of Galileo's instruments, including the incredibly tiny telescope with which he first saw the moons of Jupiter. It was amazing that such a tiny instrument, smaller than a modern spyglass, could have so changed mankind's world view. The museum seemed dark and unfrequented, filled with a jumble of scientific apparatus; some of the instruments appeared of very dubious value, but I had the impression that everything connected with science was of great interest to my mother.

On leaving Florence she took me south to Rome, stopping *en route* at the charming and rather remote mediaeval town of San Gimignano 'of the beautiful towers'. There we walked the sunny, dusty streets and prowled the tiny art gallery where she sought out old friends among the paintings of Benozzo Gozzoli. The tall, mediaeval, stone bell towers attracted me, as had the Leaning Tower, but mother, who had a great fear of heights, even to the extent of vertigo, would not allow any climbs. We compromised with a long walk in the Tuscan countryside outside the walls of the town, making for a lone cypress tree miles away. It proved too far to reach, but that walk on a hot, sunny day between fields of ancient, silvery olive trees and fragrant meadows, was a great respite for a young traveler. She understood the longings of youth for green fields and sunny skies. I kidded her unmercifully about her passion for Duccios, groaning in comic protest whenever she showed me yet another painting, but she bore my ribbing good-naturedly.

From San Gimignano she took me to Siena where a mediaeval celebration was in progress. Young men were dressed in bright mediaeval costumes, waving flags in a kind of dance to accompanying drum rolls. There we saw the imposing cathedral striped in black and white with its Moorish influence, and yet more paintings by Duccio and Cimabue. I quickly became accustomed to the tremendous amount of walking and the heavy doses of church architecture and mediaeval painting that were important parts of this Italian journey.

From the scorching streets of Rome she took me into the coolness of the Vatican Museum and St Peter's. The Pantheon she pointed out to me as an important architectural achievement. Outside Rome, she took me to the catacombs and the nearby monument to the blind martyr, Saint Cecilia, patroness of musicians. We also spent a day at Castel Gandolfo, summer residence of the Pope, where she visited Father O'Connell, an astronomer at the Vatican Observatory. He graciously drove us out to Lake Nemi in the Alban Hills where we stopped in the shade and my mother pointed out to me the little wood where Sir James Fraser sets the opening scene of his classic *The golden bough*. In this anthropological study of the origin of the myths of mankind, he describes the new aspirant stalking the old king, whose passing represents the setting of the sun or the waning of the year. The old king must die so that the new king shall live. It was a locale replete with ancient associations.

Back in Rome she took me to the classical sites, the Forum, the Colosseum, later the Sistine Chapel, and numerous churches. Perhaps her favorites were Santa Maria sopra Minerva, a Christian church which stands on the foundations of an ancient temple to Minerva, and San Clemente, in whose lowest regions, dark and dank with the drip of water, she showed me the ancient temple to Mithra, the oriental sun god who wore a Phrygian cap. She demonstrated how these antiquities traced a continuous history back to early times.

The later years

I recall the 1950s as a time when my mother continued her astronomical work and became acclaimed with her charming popular book *Stars in the making*. She was delighted with its success and this mitigated somewhat her difficult times with the Velikovsky controversy.

Velikovsky had written *Worlds in collision* (which, in ignorance, I had once called 'Worlds in collusion', a title my mother chuckled at, and said might have been more appropriate). Mother was asked to review it, which she did, debunking it. She was characteristically swift and erudite in finding specific biblical references to counter each of Velikovsky's own such references. This, and actions by other people, made a great furore, the scientists were accused of suppressing the book, and mother was made very unhappy over the whole thing. I expect it was a relief to her when the furore died down.

During the 1950s, my mother became interested in her family history. In a bequest from her aunt, she received a large box of dusty, fragile, virtually illegible letters, all thrown together higgledy-piggledy and with little or no information about their authors. The letters were on fragile, thin paper, often 'crossed' in the writing, after the habit of nineteenth-century correspondents, to save sending another sheet. The writer typically used a folded sheet, perhaps 8 × 10 inches when open, began on the first leaf and continued on the second, third and fourth. The writer would then turn the sheet 90 degrees, and continue writing, crossing what had been written before. It is possible to read such a letter, but it can be difficult.

The letters were not only crossed and faded, but also written in several languages, so it was no easy task to read them. But she tackled and subdued this task as well, preparing a series of fascinating transcriptions from the collection of several hundred letters. The letters turned out to be the large correspondence of her great-grandmother, Julia Pertz, daughter of John Garnett, and wife to Georg Heinrich Pertz, the German historian, the Samuel Eliot Morison of nineteenth-century Germany. His wife's correspondence with many illustrious personages vividly describes their tribulations as they weathered the fighting at the barricades in the revolt of 1830 in Paris, the uprisings of 1848 in Berlin, a cholera epidemic in Geneva, and the abortive invasion of General Montholon, murderer of Napoleon, in Boulogne in 1840. The letters themselves make fascinating reading and mirror events of the time that are still reflected in today's Europe.

Her examination of these letters, which she was just beginning at the end of her life, culminated in her publication of their most important passages in a small book, *The Garnett letters*, which she finished in her 79th year.

During her investigations concerning the people in the letters and the correspondents, she traveled to New Jersey to visit the home of her ancestor John Garnett. She took me along, though I was only an awkward teenager, and we saw the simple colonial home, now called the Buccleah Mansion, in New Brunswick, New Jersey. It is beautifully preserved, with some of its original wallpaper brought from France. In her quest for the origins of the Garnetts, which in typical fashion she performed with original investigations returning to their original home at Bristol in England, she also assisted a friend, Helen Heinemann, with an interest in Fanny Trollope.

Fanny Trollope had a curious and quixotic career and made a respectable living, after several family reverses, first by writing travel books, later by writing novels. Her most famous book was *Domestic manners of the Americans*, in which she made fun of the boorish uncouthness she observed in her American travels. She was the mother of Anthony Trollope, who became even more famous as a novelist than his industrious mother. My mother, in conjunction with her investigations of the Garnett letters, discovered the date of Fanny Trollope's birth, which had previously not been known. With a typical gesture, she made no claims concerning this discovery, but simply offered the information to her friend.

During the latter part of the fifties, I was away at school and saw little of my mother. During this period Shapley retired and Russell died. The appointment of a new director at the Observatory was a source of great anxiety to her, as I learned later, and I suspect the difficulties of the time broke some long-standing friendships. At about this time she received a professorship, which was a source of very great pride to her. In those days I was not attuned to the difficulties faced by a woman in the professions, particularly at Harvard, a bastion of maleness, but it seemed nevertheless a very great achievement that she had received this honor. Of course, the honor also carried with it many burdens, such as being Chairman of the Department of Astronomy. The responsibilities of the job put her under a great strain, which she combatted with cigarettes, coffee and the occasional pill.

The 1950s were also the decade when she published many books, *Stars in the making* (1952), *Variable stars and galactic structure* (1954), *Introduction to astronomy* (1954, 1956) and *The galactic novae* (1957).

This difficult period was also the time when the House Un-American Activities Committee, inspired by the invidious Joe McCarthy, was shaking a jingoistic fist at anyone it could. My father had done work for the Russian War Relief during World War II, collecting blankets and warm clothing to send overseas and both my parents had been involved in their popular Forum for International Problems. These semi-political activities marked them as 'dangerous', and my father was duly called before a traveling arm of the committee that came to Boston. This made both my parents uncomfortable. That difficult period passed, however, but the spectre of those unhappy times was hard to erradicate.

In 1958, my parents traveled to the Soviet Union to attend a meeting of the International Astronomical Union. Once they would have refused this invitation due to my father's ambiguous status, as he had left Russia under very peculiar circumstances at a critical time in her history. From his eight brothers and sisters, only two had survived the war and my parents were able to visit only one, Volodya, who lived in Moscow. They could not visit his sister, Lola, because she lived in a town that had no tourist accommodations. The reunion began well and perhaps ended not so well, as many such family reunions do, but it was an important experience for all of them. On the trip, they traveled both to Tashkent, to see the fabled city of Samarkand, and later to Yalta, where the beautiful beaches were tidy and childless. For my father, it must have seemed a strange contrast to his past life on Crimean beaches.

When it was time for them to leave for Austria, they arrived at the airport and discovered that only one plane ticket had been provided for them both. With characteristic courage, my mother said to my father, 'You take it'. She was prepared to stay, so long as he would leave. In the end, a second ticket was provided and they departed together.

It may seem that one should not make too much of this incident. It may all have been a mistake; but perhaps it was not. However, I think she was right. Her visit to Pulkova in 1933, the mysterious death of her beloved Gerasimovich and the war years, had left their mark.

In the late 1960s, she and I collaborated on the revision of her textbook, *Introduction to astronomy*. We enjoyed working together, although each did her part of the text independently. At her suggestion, we separated the book into two parts. I did the earlier chapters and she did the later ones. As there was a great deal going on in all parts of astronomy at that time, this worked out to be a fair division of labor. Since one of her specialties was variable stars, she expended a large part of her effort in describing the variables and elucidating their value to astronomical understanding. We also spent many collaborative hours discussing the illustrations, their informative captions and their placement in the text. Obtaining the photographs so essential to the book was in itself an interesting task, but her guidance and patience made it all seem straightforward. She was an exceptionally easy person to work with, helpful and patient, yet giving complete freedom to her collaborator.

A few years later, she and I also collaborated on an examination of the Celescope ultraviolet magnitudes, a task as difficult as it was disappointing. (The television cameras of the experiment orbiting in a satellite far above the earth produced inconsistent results.) Working with her was always enjoyable and edifying; it is a rare and marvelous experience to work with someone who is reasonable, informed and indefatigable.

In the early 1970s, she and I traveled together with my two boys in England. We drove about to see museums for us and castles for the boys. We visited ancient earthworks and travelled to Scotland, staying for a night in quiet Reeth, at the head of Swaledale. We also went briefly to Paris where we opened up the wonders of a crowded Louvre to the boys.

Traveling with her was enjoyable, as she was so reasonable and knew so much, but it had its pitfalls. With true Edwardian fervor, she always packed too many clothes and was reduced to lugging heavy suitcases across railway concourses at a labored pace. My son George, quick to perceive this aspect of traveling with his grandmother, stole the show by packing only a single pair of socks in his bag (he was already an experienced traveler). I suppose he decided that he would help by carrying his grandmother's bag, which he did, and if he needed clothes, we could buy him some, which we did. Thus I realized that in some things the young were quicker than I.

When we were in England, we noticed that she had a passion for English sausages. Wherever we stopped to eat in London, she always inquired for them; it seemed she could never get enough. Having tasted some myself, I could not truly understand her passion for this somewhat plebian fare, for though they do have a pleasant flavor and are quite unlike American sausages, they do not seem in any way exceptional. Perhaps, like several of her passions, it was the result of early associations. Her love of music, for example, was not a broad love of many types of music, nor an analytical sense or interest in its mathematical qualities, but a love born of associations with a particular piece heard at a particular time. Because she played the violin and attended many concerts, particularly in her teenage years, she had a large repertoire, but by and large she did not add to it. What she liked was what she knew.

These strong associations she formed so early seemed furthermore, to give a stability and imply meaning to her life. Those in whom such associations are weak or nonexistent tend to be blown hither and yon by the winds of life, but with her, associations seemed to give her a knowledge of what was right and aid her in setting goals and steering a proper course. Her love of botany, an early enthusiasm inspired when she studied under her beloved Miss Dalglish, was to serve her well throughout life. Her knowledge of wild plants was a constant source of pleasure on country walks and rambles, and she often illustrated an astronomical argument with a botanical analogue. This was particularly apparent in her Russell Lecture.

Her intense love for the theater was another such love. Extensive theater-going in London at an impressionable age gave her many experiences in this arena. She carried this love to her home by reading many plays for pleasure. She translated this love into astronomical terms when she wrote *Stars in the making*. Late in her seventies, on a solitary trip to London, after seeing yet another performance of Shakespeare's *King Henry IV, Part I*, and another of *Abelard and Heloise*, she asks herself

> Do I know all the stories in the world?
> Nothing is new or surprising anymore.
> Or is England always looking backward?
> Or have I become accustomed to looking forward?[19]

No one except possibly the participants can say what causes two people to love each other, or what causes a rift between them. Insofar as any family member can be objective, I supposed that my parents loved each other intensely at first (when they 'eloped' to New York to get married), and that the problems of bringing up a family on a very limited income (her salary was small and his smaller) in an era of difficulty must have been large. During the early years, they always managed to find a housekeeper, but later this became impossible and the household was often chaotic. There were professional disagreements, not so much on scientific issues as on who should go where when. She was more important scientifically, so she generally left him as babysitter if there were no other alternatives. He, in turn, sometimes behaved flamboyantly, which, like the stoic she was, she tried to ignore.

Difficulties there were, but like two magnets of opposite poles they stuck together. He needed protection, as he thought he had experienced under the Tsar, and she made a commitment she would not break. It was a characteristic attitude. In truth they loved each other deeply. Her last conscious thought was of him; without her, he was lost forever.

In later years, my parents traveled a great deal. They attended annual meetings of the American Astronomical Society held at many locations around the United States and, every three or four years, journeyed to the meetings of the International Astronomical Union, held in major cities throughout the world. Together, they visited many exotic places – India, Japan, Hawaii, Trinidad and Tobago, Alaska, Egypt, Spain. They also took a long tour of South America, visiting the southern observatories.

The final journey

On their final journey together in 1979, they circumnavigated the globe, flying westward to Tahiti and Australia, where they visited the friends my father had met in 1956. (He had spent a year in Canberra making observations of the southern sky.) They also stopped in the Philippines to visit Sr Francis Heyden, a long-time and most congenial friend. He had befriended me too, when I was at school, showing me the treasured first edition of Newton's *Principia* in the observatory at Georgetown University, where he had spent many years. From Australia they travelled to India and made a further stop at Istanbul, Turkey, whence my father had fled when he left the Soviet Union.

Mother was entranced to see, for the first time, the Hippodrome, where the Byzantine courtiers held their chariot races, and the marvel of Hagia Sophia, with its lofty dome. She told me that the great green marble and porphyry columns inside this stupendous edifice had come from the Temple of Diana at Ephesus, one of the Seven Wonders of the Ancient World. It seemed fitting that this woman with so much vision had seen these ancient artifacts and unraveled this connection during her last trip.

When she returned, early, from that trip, in which the last miles were filled for her with exhaustion and agony, she was clearly ill. In August she entered the hospital for a biopsy. Lung cancer was diagnosed and a regimen of radiation treatments was prescribed. She grew weaker and weaker, at first still moving from her bedroom to the first floor of her home, then confined to the second floor, finally restricted to her bedroom. A week after the radiation regimen was complete and she was pronounced cured, she was hospitalized. Stoic to the last, no moan of agony, no complaint ever crossed her lips.

She was clearly sinking. This was hard to take philosophically. When a well-meaning social worker visited us at my mother's bedside in the hospital, I lambasted her vehemently for being at least two months too late. Social work in the home during the advancement of the last weeks of her stay there would have been far more useful to my mother than a visit to the bedside of a dying woman. When I gave this tirade, my mother was lying with her eyes closed, heavily sedated. She was sleepy with heavy doses of morphine, given to relieve the pain caused by the destruction wreaked on her once strong frame. She heard what went on about her, however, for when I had concluded my diatribe she turned her face toward me, opened her eyes just slightly, whispered 'It's alright, dear' and closed her eyes again. These were the last words I heard her speak.

When we could no longer converse with her, we brought her music. My brother Edward brought Mozart's *The magic flute*, that comic opera she enjoyed so much, and my husband John brought Handel's *Messiah*. A beatific look came across her face as the first strains of that soft music reached her ears from the tiny speaker placed on her pillow. The following morning she died in her sleep. She gave her body to science and is buried in the graveyard of the Tufts Medical School in Tewksbury, Massachusetts. In the words of Rupert Brooke

> . . . some corner of a foreign field
> That is forever England.

Even after her death, she found a personal way to serve science.

K.H.
July, 1982

Notes

1 Payne, C. H., diary, 1916.
2 Struve, Otto and Zebergs, Velta, *Astronomy of the 20th century*, New York & London: Macmillan, 1962.
3 Struve, Otto, *Science*, **126**, 1350-1, 1957.
4 Pannekoek, A., *A history of astronomy*, London: George Allen & Unwin, 1961.
5 Lang, Ken and Gingerich, Owen, *Source book in astronomy and astrophysics*, Cambridge, MA: Harvard Press, 1979.
6 Öpik, Ernst, Obituary, *Irish Astronomical Journal*, **14**, 69, 1979.
7 Hoffleit, Dorrit, The end of an era, Cecilia Payne-Gaposchkin and her last book, *Journal of the A.A.V.S.O.*, **8**, (2), 48-51, 1979.
8 Whitney, C. A., Cecilia Payne-Gaposchkin: an astronomer's astronomer, *Sky and Telescope*, March, 212-14, 1980.
9 Mihalas, D., letter to G. Field, 18 December 1979.
10 Morgan, W. W., letter to Katherine Haramundanis, 27 August 1982.
11 Smith, Elske v. P., Cecilia Payne-Gaposchkin, [obituary], *Physics Today*, June, 64-5, 1980.
12 Payne, Cecilia Helena, 'The journey', July-September 1911.
13 Payne, C. H., diary, 1916.
14 Powell, Dilys, 'The traveller's journey is done', London: Hodder and Stoughton, 1943.
15 See 'The dyer's hand', Chapter 17.
16 Payne, C. H., diary, 1933.
17 *Ibid.*
18 Eddington, A. S., letter to Cecilia Payne, 17 October 1933.
19 Payne-Gaposchkin, Cecilia, diary, 1974.

Acknowledgments

In preparing this memoir I have received help and encouragement from many who loved and admired its subject. I owe a debt of gratitude to several individuals, among them Charlotte Sitterly, Vera Rubin, Martha Liller, Jesse Greenstein, Charles Whitney, George Field and Barbara Welther. I am particularly grateful to early readers of the manuscript, to Dorrit Hoffleit and my husband John for their advice and timely suggestions, and to my sons George and Sergei for their perceptive criticisms.

The dyer's hand
an autobiography

Copyright 1979 Cecilia Payne-Gaposchkin

Dedicated to the
Tres Matrones

Elizabeth Edwards
Dorothy Dalglish
Ivy Pendlebury

Foreword

The short account on the next two pages was written by a close school friend of Cecilia when they were both at Newnham College together. The school friend, Betty Grierson Leaf, the niece of Walter Leaf, died tragically about 10 years after the writing of this description. As Cecilia neared the end of her life, she asked that this all too brief text be used as her obituary notice. In accordance with her request, it is included here. The account was printed by R. I. Severs, Hobson Passage, Cambridge, England, in 'Thersites' on Saturday, June 2nd, 1923. Cecilia died on December 7th, 1979.

Katherine Haramundanis
Lexington, Massachusetts
December, 1979

Cecilia Payne

At the age of five Cecilia saw a meteor, and thereupon decided to be an Astronomer. She remarked that she must begin quickly, in case there should be no research left when she grew up.

She generally accomplishes what she has determined to do, and in spite of some years spent in other pursuits, a lecture on Relativity has recently brought her back to her first love. Though her epoch-making discoveries are yet to come, she is now to be found on starry nights among the moths and dust of Newnham Observatory, showing the wonders of the Heavens to all who come. She practices her powers of exposition in addresses to the Science Society (on these occasions she is President, Lecturer and Lantern-boy all in one), and somehow conveys a glimmer of her own enthusiasm to the most unimaginative.

She is most completely happy when some 'beautiful' mathematical theory of the Universe makes her forget the minor disturbances of everyday life, though this absorption in abstract questions is sometime fraught with danger. I have known her suddenly descend from her bicycle amid a Saturday crowd in Petty Cury, and unaware of the whirlpool of vehicles, abruptly enquire 'What's the good of thinking?'. Nevertheless she is always thinking of *something*, and when safely lying on her back on the floor (she despises armchairs), she will talk of all things under the sun, from ethics to a new theory of making cocoa. She combines a tendency to speculation with rigorous honesty of mind; this is her one touch of severity – and it is chiefly exercised upon herself. Though she refuses to go to bed, until any stray thoughts provoked by some chance discussion have been properly classified and pigeon-holed, her interest in material tidiness is chiefly theoretical, and she has a feeling of slight surprise when she loses an important letter which she has 'been carefully keeping on the floor'.

In spite of her academic triumphs, Cecilia is not an intellectual ogre. Though she is at home among the stars, she is none the less human and good to live with. She has a quick way of appreciating people's qualities, and is essentially sociable. All' Hallowe'en and birthdays are the occasions of festive gatherings in her room, and the more guests she can then squeeze into her room the better pleased she is. In her lighter moments she is addicted to puns, and having once acted the part of Richard III, she has since had a suppressed craving to play the villain in melodrama. Crime, however, is not part of her daily life. She has been known to rescue a slug from the salad, and it weighs on her conscience if she accidentally swallows a fly. In spite of traits such as these, which recall the 'Nonne-Prioresse', she labours under the delusion that she should have been a man, but I suspect that those who know her as she is would hardly wish her otherwise.

Perhaps the most surprising thing about Cecilia is her many-sidedness. She works off the effects of an overdose of mathematics by vigorously conducting the orchestra or hall part-singers, and during the miseries of influenza she sits up in bed with her fiddle, and declares she has been spending the morning at the opera. Books too, are her familiar companions. *Isaiah, Hamlet, Punch* (since 1880), 'The hunting of the Snark' she knows by heart, and Plato will soon be added to these. When even books and music fail, she takes refuge in the construction of mechanical toys, and a confusion of cardboard and cotton will soon turn into a figure of Icarus, who waves expressive legs, arms and wings.

Cecilia's activities but half describe her. One can only say of her as of her favorite literary 'hero':

These are actions that a man might play, But 'she' hath that within which passeth show.

PART I
The vision splendid

1

Backgrounds

I spring quite literally from a pagan background. The name of my Father's family was Payne – derived, so he said, from *paganus*, the word applied by the Roman conqueror to the aborigines of Britannia, comparable to the Greek 'Barbarian'. There have been Paynes at the foot of the Buckinghamshire Chilterns, where I was born, for nearly a thousand years. In the Domesday Book, Payne appears repeatedly in the lists of Buckinghamshire owners of land. 'In Hochestone (just north of Aylesbury) Payn hold eight hides and two virgates of William. There is land for ten ploughs . . .' I suppose Paynes have tilled the soil there for at least ten centuries, down to my grandfather who had a farm near High Wycombe. My roots are deep in the soil of the English Chilterns; no other countryside looks, feels, or smells like home. My Father, having become a lawyer, a musician and a scholar, left his Oxford Fellowship late in life, to marry and bring up his family in the village of Wendover on the native soil of his ancestors. I was born there in 1900.

But I am far from being 'mere English'. Traditions of many nations have converged in me and mine. My Mother was German by nationality, herself descended from two generations of German–American and German–English unions. My Husband was born in Russia. I have a Greek son-in-law, an English–French daughter-in-law. The international spirit, the urge to range over the world, is in my blood. I left my native land to make a life in the United States.

Already in the eighteenth century my Mother's forbears were moving about the world. I can look back on a great-great-great-grandfather whose two sons left their native Hanover for Russia in about 1760. The elder, a goldsmith, married and settled there. His

Fig. 1. Emma (Pertz) Payne (mother)

younger brother Christian August Pertz was called to Moscow at about the same time to establish a bookbinding business. It was the time when Catherine II of Russia was drawing on the services of German craftsmen, and he told his children that he would have been happy to make his home in Moscow if it had not been for his old Mother in Hanover, whither he returned in about 1790. Georg

Heinrich Pertz, the historian, my great-grandfather, was born there in 1795.

Another great-great-grandfather, John Garnett, left his native England for the young United States in 1795. He was a prosperous merchant, but he was disenchanted with England. The immediate cause of his decision is said to have been the popular persecution of Joseph Priestley. He chose the life of a gentleman farmer in New Jersey, and there he brought up a son and four daughters as American citizens. After his death his widow and three daughters returned to Europe and made their home in France. His daughter Julia became the wife of Georg Heinrich Pertz, and provided me with great grandparents.

At the time of his marriage, Georg Heinrich was Librarian to the King of Hanover, but later was called to Berlin as Royal Librarian, and undertook the publication of the *Monumenta Germanica*. His wife's letters give a vivid picture of the period, for she was a close friend of Lafayette, Fanny Wright, Fanny Trollope, the Martineaus and the historian Sismondi. Her two sisters, who never married, are shadowy figures whom my Mother knew as a child, a pair of old spinsters living in Brighton. The only material relic that I possess of the great fortune that John Garnett carried from England to New Jersey is a little rhinestone brooch.

Julia gave her husband three sons of whom Hermann, the youngest, was my grandfather. Soon after her death, her husband (now the Chevalier Pertz) met with Leonard Horner, the Scottish geologist, father of six beautiful and accomplished daughters: Katherine, Susan, Mary, Frances, Leonora and Joanna. They must have been a redoubtable brood. The story has come down that once, while guests at a country house, they were distressed (rock-ribbed Presbyterians that they were) at having nothing useful to do. So they tore all the edges off their handkerchiefs and re-hemmed them.

Leonora became the second wife of Georg Heinrich Pertz, endowing me with a step-great-grandmother and five step-great-great aunts. Most of the sisters lived to be over 90, and two of them are vivid memories of my childhood. Mary became the wife of Sir Charles Lyell, Katherine married his brother, Captain Lyell, and Frances was the wife of Sir Charles Bunbury. Susan and Joanna were unmarried;

they lived in Florence for many years and were the authors of *Walks in Florence*, a book well reputed in its day.

Aunt Katherine Lyell and Aunt Joanna Horner lived in London, and we often visited them as children. A visit to Aunt Katherine was an alarming and delightful adventure. The great Sir Charles Lyell had been her brother-in-law and she had known Darwin and Huxley, whom I worshipped in secret; but she was such a formidable old lady that I never dared to ask her about them, much as I longed to do so. One spoke to her as one addressed Royalty: she chose the subject. But she did give me a few of their letters. I can remember vividly how she went to the piano and accompanied herself as she sang: *Oh dear, what can the matter be?* for the entertainment of us children. She was 97, I think, at the time.

Aunt Joanna Horner was sweeter and gentler. Like her sister, she wore a black silk dress with a lace fichu and a lovely little white lace cap. She used to talk about the past, the time when there were no railways and one traveled by coach. I remember her punctilious early-Victorian speech: she always said 'ospital' and 'otel', though I do not recall that she ever said 'obleeged'. Like Great-great-aunt Katherine, she used to burst into cheerful song. I can remember

> Elijah was a Good Old Man
> Who went to Heaven in a fiery van.
(Chorus)
> Then let us all be a Good Old Man
> And we'll go to Heaven in a fiery van.

The next generation provided me with four great-aunts. Hermann Pertz, Julia's youngest son, married Emma Marsh Wilkinson, daughter of the formidable James John Garth Wilkinson, who is a story in himself. He was a Durham man, son of a judge who seems to have been at least as formidable as his son. There is a tale that the judge came upon an old lady who was hesitating to cross a miry road. He picked her up bodily and carried her across, whereupon she turned on him and berated him soundly. So he picked her up, carried her back, and deposited her on the other side again.

There was a tendency of later Wilkinsons to associate themselves with the John Wilkinson who was involved in the construction of the first iron bridge and of the pioneer ironclad ships. But I suspect that the relationship, if any, is a very distant one. At the end of the eighteenth century there was a great social gulf between a judge and an ironmaster.

Great-grandfather J. J. G. Wilkinson became a surgeon, and had many interesting patients, including Lola Montez, Augustus de Morgan, and the great actress Geneviève Ward; but his heart was in many other fields. He was a close friend of Emerson and of William James the Elder. He it was who issued the first printed version of Blake's *Songs of innocence*, and, though he was the fierce foe of the nascent cult of spiritualism, he was a convinced mystic. He embraced the tenets of Swedenborg, and edited editions of that visionary's works in English. He produced, too, an amazing volume of poetry executed by automatic writing, *Improvisations of the spirit*, which cast a strong influence over my early years. He was, for the time, quite a traveller. In 1869 he visited the United States, and the diary that he kept on that journey is still full of interest. In 1868 he made an even more unusual visit to Iceland; he had a profound interest in ancient Icelandic literature, which has stimulated me in my turn to a study of the Edda.

J.J.G.W. had three daughters and one son. Emma was married (at the age of 16) to Hermann Pertz, and thus I acquired a grandmother. My grandfather was an engineer in the Prussian army, sweet and gentle, but not a success in the worldly sense. The two other sisters made rich marriages: Florence to St John Attwood-Mathews and Mary to Francis Claughton Mathews. Neither had children, and their preoccupation with material wealth gave me an abiding horror of money, and a feeling that there is something cruel and indecent about being rich. For they used their wealth in an attempt to reduce the descendants of their less-fortunate sister to subservience.

I can see them now: horribly stout, clad in black satin, enthroned like Far Eastern idols, and fondling obese little pug dogs. They reigned over their estates, Aunt Florence in Monmouthshire, Aunt Mary in Hampshire, tyranizing over unhappy, obsequious 'companions'. I paid (by 'Royal Command') one brief terrified visit to Great-aunt Florence. Her domain, Llanvihangel Court, was the grander of the two, complete with the Bed in which Queen Elizabeth

Slept and the Bloody Footprint left on the stairs after a duel. From her awful eminence she recommended that I should marry an elderly bishop, a distant cousin of her late husband. I had, of course, never met the bishop. I told her stoutly that I was perfectly capable of earning my own living; I was 17. That was the last time I saw or heard of Aunt Florence Mathews. She and her sister lived to great ages, and died unlamented. These were my full great-aunts.

How different were my step great-aunts, daughters of G. H. Pertz by his second wife, Leonora Horner! His first marriage had brought him three sons; his second produced three daughters – Annie, Elizabeth and Dorothea. Elizabeth died in childhood; Aunt Annie and Aunt Dora were the beloved relatives of my early years. They lived together in Cambridge after their parents died, and every year we used to spend a week with them. They had inherited the stern rectitude that had impelled their mother and her sisters to re-hem their handkerchiefs rather than sit with their hands before them. Aunt Annie was an accomplished painter. Aunt Dora had been one of the early students at Newnham College and had been trained as a botanist. It was from her that I conceived the hope of becoming a scientist some day. It was characteristic of her that she never spoke of her own work; I believe she assisted Sir Francis Darwin in his research on plant physiology but I could never get her to tell me much about it. I know that she inspired me with the dream of being a student at Newnham one day, but she inspired by example, rather than by precept.

Emma Wilkinson, married to Hermann Pertz, had two daughters and two sons: Emma, Florence, Fritz and Eddie. Emma married Edward Payne, and thus I acquired a Mother. Her sister Florence was my only aunt – a brilliant outspoken woman who became a professional pianist. She had no patience with mediocrity and did not suffer fools gladly. I loved her, but I feared her too. Her unsparing criticism made me stop trying to play the piano, but she fostered my love of music and took me to many a concert. In her company I heard d'Albert and Carreño, Busoni and (at the end of her life) von Karajan, and from her I learned to know the Beethoven piano sonatas and the piano concertos, and the multifarious works of Schumann. She played, as I recall, magnificently. Alas, her flowering came at the time of the First World War, when a combination of German ancestry and the name Pertz were enough to close the doors of the concert hall to her. She remained a brilliant and embittered soul.

Aunt Florence ('Flossie' as we called her) had a fascinating social circle. At her parties one would meet Lord Buckmaster (the Lord Chancellor), the Master of Chancery, Wedgwood Benn, Sir Israel Gollancz, Etheldreda Hull and many more. Her most valued friend (though he did not appear at her parties) was Henry James, whose father had been intimate with Dr Wilkinson. I grieve to admit that Henry James was thrust down my throat to such an extent that to this day I cannot bring myself to read any of his books – to my great loss.

Women dominate my background: memory recalls great-great aunts, great-aunts and an aunt. Few of them married. There is hardly an uncle in the picture. Henry Garnett, shadowy great-great uncle, died young, unmarried, and apparently unlamented. The two brothers of my maternal grandfather were never mentioned, save to say that Uncle George was addicted to terrible pranks, and Uncle Karl did not amount to anything. Of my Mother's two brothers, Eddie died as a boy and Fritz became an officer in the Prussian army. One of the griefs of my adolescence, during the First World War, was the fact that my only relative in that conflict was fighting on the other side. He and his were cut off from me by the tragedy of history.

This was my heritage: an international background shot through with the urge to migration. It is dominated by women in each generation, including my own, for I lost my Father when I was four years old. Of my relatives on his side of the family I know almost nothing. Life brought me no material possessions, only the world of Nature and the world of Books. I am trying to trace the path into which my heritage has impelled me.

2
Beginnings

The Bee Orchis was growing in the long grass of the orchard, an insect turned to a blossom nestled in a purple star. Instantly I knew it for what it was. My Mother had told me of the Riviera – trapdoor spiders and mimosa and orchids, and I was dazzled by a flash of recognition. For the first time I knew the leaping of the heart, the sudden enlightenment, that were to become my passion. I think my life as a scientist began at that moment. I must have been about eight years old. More than 70 years have passed since then, and the long garnering and sifting has been spurred by the hope of such another revelation. I have not hoped in vain. These moments are rare, and they come without warning, on 'days to be marked with a white stone'. They are the ineffable reward of him who scans the face of Nature.

My first sight of the spectrum of Gamma Velorum, the realization that planetary nebulae are expanding and not rotating, the fact that U Gruis and RY Scuti are eclipsing stars, the true nature of T Ceti, the period of AE Aquarii, the bright-line nature of the supernova spectrum, these are some of the moments of ecstasy that I treasure in retrospect.

And yet none of these revelations was original. Agnes Clerke had described the spectrum of Gamma Velorum before I was born. Ejnar Hertzsprung, leaning over my shoulder and putting his finger on a wrong subtraction, solved the mystery of U Gruis. The nature of RY Scuti was my Husband's discovery, not mine. I could never have understood T Ceti without de Sitter's photographs, nor AE Aquarii without Alfred Joy's velocities and Merle Walker's hint. The spectrum of the supernova had been surmised to be in emission before my day, but it was only a surmise. In a moment of enlightenment I *knew*. As I looked at the famous spectrum of Z Centauri (classified at Harvard many years before as a carbon absorption star), I suddenly saw it reverse, as one sees the craters of the Moon suddenly turn into hillocks, and absorption translated itself into emission. There is nothing personal in the thunderclap of understanding. The lightning that releases it comes from outside oneself.

I did not pluck the Bee Orchis. I stood and gazed at it; I see it now, rising in the long grass under the apple and hazel trees. I ran into the house to bring the news, and my Mother could not believe me. I must be mistaken: such a flower could not be growing in our homely Buckinghamshire soil – here she was mistaken; it is a characteristic plant of the chalk – but when she saw it she was convinced. Deering, the old gardener, was called to transplant it to the garden under the little spruce tree.

Orchis and spruce tree became a little shrine. I made a secret cult of the transplanted Christmas tree, and I well remember standing before it and taking a vow to devote myself to the study of Nature. Trees were my earliest companions. A little grove of maples had been planted at the bottom of the garden and there I used to spend my days with my Unseen Playmate, whom I called Grenson. The eldest child of my parents, I had no other, and he was very real to me. The maple grove was always 'Grenson's Wood' to me, and within it I wove a childish mythology of the world around me.

Nature was very close. After a summer rainstorm I found the garden crawling with black slugs, and wept bitterly that the world could contain anything so horribly repulsive. Winter brought rare, beautiful snowstorms. I shall never forget my first encounter with snow. My Mother was wheeling me in my pram – I must have been about three years old – and I looked at the lovely, fluffy white carpet. It must be as warm and caressing as my swansdown pelisse. Nothing would satisfy me but to feel it round my feet. At last (in exasperation, I fear, for I was a most determined child) my Mother lifted me down. Instead of the gentle warmth I had pictured, I felt an icy chill. How I wept, and how tenderly she took off my shoes and rubbed my little feet! When I was eight years old there was a heavy snowfall in April, and the blooming daffodils were buried beneath a white blanket. I was so terrified that I became quite ill, and I still experience a feeling of terror when the snow begins to fall.

One winter evening my Mother was wheeling me in my pram, and we saw a brilliant meteorite blaze across the sky above Boddington Wood. She told me what it was, and taught me the right name for it by making a little rhyme

> As we were walking home that night
> We saw a shining meteorite.

It was my first encounter with astronomy. Soon I learned to look for 'Charles' Wain' in the northern sky, and to recognize Orion's belt. When I was 10 years old I was taken out to look at Halley's Comet, a most disappointing sight! It could not compare with the great 'Daylight Comet' that had blazed its tail across the western sky earlier in the year, a truly awesome experience.

It did not take me long to feel that I lived in a man's world. A brother and a sister joined me in the nursery. I felt from the first that my brother was the one who really mattered. My first reaction to my brilliant, warm-hearted brother Humfry was one of jealousy. There came a day when his godfather drove up in a carriage with two horses. He invited his godson to go for a ride in that glorious equipage. In vain did I beg to be taken too. To add insult to injury, Humfry was allowed to be 'the red-legged partridge', to wear the coveted red gaiters that were a mark of special distinction. I watched them canter away, and it seemed like the end of the world. I wept inconsolably; I ran for comfort to my Father, who was sitting in his study writing his *History of the new world called America*. I shall never forget how he laid aside his pen and turned to me. 'Never mind, Popsy,' he said, 'I'll take you for a walk,' and he did. I think he must have known which way to go, for presently the two horses came in sight, and he signed to the coachman to stop. I was lifted up, and rode behind those two steeds into the seventh heaven.

What a godlike figure my Father was! Every evening we used to watch for his return from London, where he had a legal practice. He came home over the fields in winter, swinging an enormous lantern with a tallow candle in it. He would bring us chunks of maple sugar, and would sit with us on the nursery floor, building 'rude stone

Fig. 2. Edward John Payne (father)

monuments', miniature Stonehenges, with the wooden bricks that he had had specially made for us.

He was a gifted musician and the house was full of music in my earliest years. When I was two weeks old he began to play scales to me on the recorder, 'to educate your ear'. I think he must have succeeded only too well, for I have perfect pitch, and such a fastidious

Fig. 3. Cecilia Helena Payne 1904

ear that most singers and many string players seem to me woefully out of tune. One night I crept downstairs from my bed and peeped into the room where he was playing the 'cello in a string trio. I remember the rhythm of the triplets – I went about singing the music afterwards. One of his treasures was a small eighteenth-century pipe organ. He taught me to sing the 'Doxology' and *O Sanctissima* as he played it. I can still hear him singing and playing *The Galway piper* and *The British grenadiers* (which I thought was a song about deer going over a bridge). These are the happiest memories of my childhood. I inherited his love of music and his beautiful violin.

When I was four years old, my Father died suddenly. I shall say nothing of that traumatic experience, which I remember only too well. My Mother could never afterwards speak of him without tears. For the rest of my childhood I felt I was not like other children, for I had *two* fathers in heaven.

3

Prelude to education

Isolated from the world, we grew up in an atmosphere of literature and art. Our first story-book was the *Odyssey*. The tale of the return of Odysseus, and his recognition by the faithful Argus, still moves me as it did when I first heard it at my mother's knee. Jason and Perseus, Theseus and Heracles were the figures of our folklore. My Mother, an accomplished painter, surrounded us with beautiful pictures. The Dresden Madonna hung above the fireplace in the nursery.

Our tastes were not passively acquired. Well do I remember, on a visit to the village grocery, being given a large colored advertisement for boot polish, a picture of a rubicund policeman in roars of laughter. We thought it was beautiful, and climbed up to fit it into the frame of the Dresden Madonna. I can still see that jovial face – a contemporary 'Laughing Cavalier', but our exhibit, understandably, had short shrift.

When I was a small child I used to ask to be told a story after being put to bed. Perhaps the supply of stories ran low, or the narrators wearied, and I was given to understand that the practice must cease. 'Very well,' said I, 'I will tell stories to myself.' And I did so, making them up aloud as I went along. For years I was my own audience, and later I would compose and recite a nightly story for my sister.

When I learned to write it was a small step to set the stories down on paper. I have before me the manuscript of what was, perhaps, my earliest literary effort, laboriously written in pencil in an exercise book. I must have been about six years old. It is a curious commentary on a child's psychology, and shows that by then I had been exposed to Grimm's Fairy Tales. I reproduce the spelling and punctuation exactly

once upon a time there was a king and queen I am sorry to say she was very naughty. one day she went at night she went to an inn and with the kings sword she went. wel wen she got there she asked if she cod go in to the bedrooms and wen she got there she cut of there heads and there bodys were swiming in blud then she run away. wen the innkeeper came up he was extremley angrey. he sent for the ploice officers. this is the ending of it

I am glad that my later writing has been less sanguinary. Could the author of those lines have had the makings of a pacifist?

I had been introduced to education at the age of about six. By a piece of extraordinary good fortune, a small school was opened, just across the street, by Elizabeth Edwards. She was Welsh by descent and education, and had a remarkable gift for teaching. 'A good education,' she used to say, 'will make you do a thing you do not like, willingly and well.'

The first joy of school was learning to read. 'Shall I be able,' I remember asking, 'to read the *Encyclopaedia Britannica*?' Assured that I should be able to read *anything*, I fell to with a will. I remember the pages of the primer as if it were yesterday – the vowels, labials, gutterals, sibilants and silent letters printed in different colors, red, black, blue, green and yellow. Reading became my joy and relaxation, and has been so ever since. Our house was full of books, not only my Father's library, but legacies from many forbears – Latin, Greek, French, German, Spanish, Icelandic books, waiting to be explored. I was fascinated with words, eager to put them to use. I recall a tea-party at which I was to hand round the little cakes. 'This one,' I indicated to a special guest, 'is unique' (making the word rhyme with curlicue), and was greeted with a roar of laughter, to my extreme chagrin. I must have picked up the word from a book, and discovered its meaning, but not its sound.

At school we were trained in memory and observation. We seldom used books, and even the long poems that we learned to recite in unison were taught by word of mouth; *The Pied Piper of Hamelin* and *Hervé Riel* were among them. I recall that I repeated the final lines of the latter poem, beginning 'Name and deed alike are lost' word for word after hearing them read just once. A very good memory has always been one of my assets, and I believe that I owe it entirely to this early training.

Another valuable discipline was the incessant drill in mental calculation. The British coinage (now, alas, a thing of the past) provided fascinating daily exercises, in which we responded to: 'What is the cost of a dozen articles at eight-and-sixpence-three-farthings each?' with lightning speed – or went to the bottom of the class. Decimal coinage, despite practical advantages, offers no such delights.

At school I discovered to my cost that I was left-handed. No concessions were made to this weakness, and I remember the physical discomfort, almost amounting to pain, that it gave me to hold the pencil in my right hand. It cost me an effort – indeed it still does – to distinguish left from right, and was an agony in dancing class, where my height placed me at the head of the line, and there was always confusion as to which foot should move first. Many years later, in my teens, I happened on a pamphlet that advocated the development of both hands, written by my redoubtable great-grandfather Garth Wilkinson. The exercises that he described involved the practice of mirror drawing and writing, using both hands at the same time. I practised them religiously, and became essentially ambidextrous, with the additional accomplishments of writing backwards, upside-down, and upside-down-backwards. The only thing I am unable to do today with my right hand is deal cards: nobody thought to break me of doing that with my left.

School discipline was strict. When we went for a walk, we marched. During our oral lessons we sat with backboards (thin slats of wood placed behind the shoulders and grasped with the hands). I think we enjoyed the military discipline rather than otherwise, and referred to Miss Edwards affectionately as 'Tom', short for 'Tommy Atkins', the archetypal British soldier. Our greatest pride was a miniature Union Jack pinned to the chest, and our severest punishment its temporary removal. We were taught to glory in the British Empire and to scorn the 'little Englanders' (even though we did not know exactly what they were).

School began every day with the singing of a hymn. The attitude to religion was respectful and matter-of-fact, and if there was any religious teaching I do not recall it. Occasionally the hymn was replaced by a patriotic song. Somehow this did not seem right to me and to sing 'What can I do for England, That does so much for me?' seemed a sort of sacrilege. I began to substitute 'Jesus' for 'England', and things went well for the first verse, but in later verses it led to all sorts of untenable statements. For the first time I wrestled privately with the conflict between nationalism and religion.

The sense of living in a man's world continued to oppress me. 'Why,' I asked Miss Edwards, 'was Jesus a man and not a woman?' 'Because in his day,' she replied, 'a woman could not have done the things he had to do.' But I did not find this convincing. She used to insist that women are 'the stronger sex', and was certainly a feminist at heart. When after my repeated questions, my Mother finally told me where babies came from, my response was: 'What are men for, then?' I do not remember what she answered. I only worked it out for myself many years later when I was studying the pollination of the cycads.

In six years that I spent at her school, Miss Edwards gave me a rich education. I sometimes think she taught me all I needed to know. At 12 I could speak French and German, had a basic knowledge of Latin and a full command of arithmetic. Geometry and algebra were part of our studies, and I delighted especially in the solution of quadratic equations. Even more important, we had been taught the principles of accurate measurement. The one piece of fine equipment in the school was a chemical balance, which we learned to use with precision and reverence. This is the program of education that I should choose for the young: languages and mathematics and a sense of accuracy. Besides these things, we had received memory training, and our powers of observation had been sharpened to a fine point. Once a week we were required to pick out with our eyes (but not to touch or reveal) three brass carpet tacks placed somewhere about the school garden.

There were blind spots in this scheme of education, too. Music was one of them; I think that 'Tom' was quite insensitive to it. And music already meant a great deal to me. When I was about 12 it fell to me to perform (in a piano duet) an arrangement of the First Movement of Mozart's Symphony in E flat. I asked permission to say something before sitting down to the piano, and when this was allowed, I said 'This is so beautiful it ought never to be played'. At about this time, too, I was allowed to play one of the simplest of Bach's Two-part Inventions on a similar occasion. I played it with delight and reverence, and waited for the expected praise. 'Very pretty, dear,' said Miss Edwards gently. I perceived that some things were hidden, even from her.

For six years, the little school at Wendover filled all my life. When I was 12, my Mother decided to move to London, particularly for the sake of my brother Humfry, who must be prepared for Public School. Miss Edwards wrote me a letter of farewell and advice. 'You will always be hampered,' she said, 'by your quick power of apprehension.' She was one of the wisest people I have ever known. I cannot count how often those words have served as a warning against hastily jumping to conclusions.

There are some memories of life at Wendover that evoke a forgotten past. Such were the great summer parties for children that belonged to the Edwardian social scene. Sir Lasenby and Lady Liberty entertained the neighborhood with a lavish garden fête at their country house at The Lee. We were taken to such another garden party for children at Chequers Court, long before it became the residence of the Prime Minister.

Most clearly do I remember a children's party given by H. W. Massingham, the famed and much criticized editor of *The Nation*. Two of his sons were our school fellows, and the whole school was invited. It was a wonderful party – Miss Edwards had to remind us that, as ladies and gentlemen, we must restrain our appetites, and under no circumstances eat more than one jelly. The *pièce de résistance*, however, was a Punch and Judy show. I had never seen this ancient entertainment before, and I watched in utter horror as Punch belabored his wife Judy, and finally hanged the Hangman. As we left, I joined the procession that dutifully told our host: 'Thank you for the party. I enjoyed it very much.' I can see Mr Massingham now, his pale ascetic face set off by red hair and moustache. He looked me keenly in the eye. 'You didn't *look* as if you were enjoying yourself,' he commented drily.

Parting from Wendover was a tragic experience. Who could wish to leave the fields and hedgerows and rolling hills for paved streets and thousands of houses and a smoky sky? My roots were deep in the Buckinghamshire soil, and at first I thought that the uprooting would kill me. Indeed, when I return to my native village, it still seems to me the loveliest in the world.

4

Birth of a dream

New vistas opened up before me. London became my second home. It had always been a brooding shadow on the horizon. My Father had gone there regularly to his Chambers in Lincoln's Inn. My Mother, too, spent several days a week painting at the National and Tate Galleries. She was an accomplished copyist, of whom my Father said: 'Turner improved upon Nature, and my wife improves upon Turner.'

While we lived at Wendover, she gave me a hoard of omnibus tickets to play with. It was the day of horse buses, and the tickets for different routes were distinctively colored – red and white, green and blue and yellow – strips of paper bearing a list of fare stops. On examination, different bus lines proved to have stops in common, such as Oxford Circus, Marble Arch and Piccadilly Circus. It occurred to me that I had the material for the reconstruction of the map of London. I set to work to solve this problem in topology, and produced an intricate network of routes; but the map was a failure: it presented the impossibility of Liverpool Street in two opposite corners.

The real London was not like the map, except in size and complexity. There were endless vistas of houses, but they were not treeless. We actually had a tiny garden that gave on a public square, but it was hard underfoot and our flower garden was a grimy little rockery. From the topmost window of our Bayswater house I could see, in the far distance, the church spire of Harrow-on-the-Hill. I used to gaze at it and think how it lay at the edge of the countryside – the green fields, the hedges, the primroses and cowslips and singing birds. The Parks seemed a pallid substitute, disciplined, dull, grimly respectable. I ached for the open fields: nostalgia is not reserved for the old.

We exchanged the free enquiring spirit of our country school for a stronghold of the Church of England, large, highly organized, and very religious. The school had its own Chapel, with services at the beginning and end of the day. There was daily instruction in the Bible, the Catechism, and the history of the Christian Church. I felt the atmosphere to be stifling from the first, and these observances had little effect in bending an unregenerate twig. I did succeed in getting excused from the Chapel by fainting during the service, a technique that I had already discovered in connection with attendance at Church in Wendover.

I am glad to have been exposed so thoroughly to the Old and New Testaments, for they are great literature, and rich sources of ancient history. I am constantly frustrated by a generation on which a biblical reference or quotation is lost. Too many of those who quote from the Bible treat it literally as the unquestionable Word of God; the others seem to pass it by. Even more inexcusable are those who quote the Scriptures incorrectly. My faith in the ministers of the Church was sadly shaken at Sunday School, when I heard one of them make a terrible mess of the story of the Good Samaritan; he ruined the point by making the *victim* a Samaritan. It has taken me many decades to overcome the resentments and doubts engendered in a well-meaning but bigoted atmosphere.

A great deal was said to us on the subject of Prayer. Finally I resolved on a practical test. When examination time came, I divided the examinations into two groups; I prayed for success in the one and not in the other. The comparison would involve numbers, and numbers were a thing I understood. The result: I obtained better marks in the group for which I had not asked for divine help. I fancy that this may be ascribed to a wish to see Prayer defeated, which had governed my grouping of the examinations. Years later I noted a possible parallel in the result of a card-guessing experiment, in which my score was impossibly *low*. However that may be, I drew my own conclusions from my experiment with Prayer. The only legitimate request to God is for courage. This conviction has persisted through the years.

Learning at the new school was not the delight it had been under Miss Edwards. Mathematics was the subject I loved best, and it was a deep disappointment. I was placed in a class that still had to grapple with long division, and algebra was still a year away (alas, for my beloved quadratic equations!). We learned French and Latin, but not German (another sore point). And in the first year there were no science lessons at all; with so many hours devoted to religion, something had to go.

And I knew, as I had always known, that I wanted to be a scientist. I resolved to concentrate on the studies that would help me to reach that goal. In the atmosphere of that school it was uphill work. Nor did I find much help at home. My forbears had been historians, not scientists, and though we had thousands of books, few were devoted to science. At last I found two that helped to fill the void. One was an old treatise on botany, using the Linnean System of classification, with text in German and French. The other was Newton's *Principia*. With the aid of a dictionary I laboriously translated the botany into English, under the impression that its contents were up-to-date, and I absorbed the Propositions of the *Principia*, though of course I could not follow the proofs. Here were beliefs that I could accept wholeheartedly.

Two other groups of books offered some help. One was the works of Emmanuel Swedenborg, especially the volume entitled *Chemistry, Physics, Philosophy*, which gave me a mystical view of Science that I never lost. The other was the collected essays of Thomas Huxley, a complimentary copy (sent to my Father?), decidedly out of place *dans cette galère*. Huxley quickly became one of my idols. I still read his essays periodically. If I learned to develop the spirit of a scientist, it is largely due to his influence.

Incredible as it may seem in the atomic age, the neglect of science teaching in a Church school in 1912 was a reflection of the nineteenth-century feeling that science is in conflict with religion. At the time when I left this school, openly resolved to be a scientist, the Principal, a dedicated, almost saintly Churchwoman, told me that I was 'prostituting my gifts'. She could not have used stronger or more shocking language. She was absolutely sincere. To her the Chapel was more important than the laboratory.

But by this time I had, in a sense, converted the laboratory into a chapel. On the top floor of the school (a town house, high and narrow) was a room set aside for the little science teaching conceded to the upper classes. The chemicals were ranged in bottles round the walls. I used to steal up there by myself (indeed I still do it in dreams) and sit conducting a little worship service of my own, adoring the chemical elements. Here were the warp and woof of the world, a world that was later to expand into a Universe. As yet I had caught but few glimpses of it – the meteorite, Halley's Comet, the Daylight Comet of 1910. I had yet to realize that the heavenly bodies were within my reach. But the chemical elements were the stuff of the world. Nature was as great and impressive to me as it had seemed when I stood under the spruce tree and vowed myself to its service. It overshadowed everything. I was appalled at the words of the hymn

> Were the whole realm of Nature mine
>   That were an offering far too small.
> Love so amazing, so divine
>   Demands my soul, my life, my all.

I could not even think of my puny, unimportant, negligible self in comparison with that magnificent panorama. The very thought was a sacrilege. I could not bring myself to utter the words.

In the second year of school things grew brighter. We began to study algebra and Euclid. And for the first time I found a friend and mentor. Dorothy Dalglish came to the school to teach science. A radiant personality, an inspired teacher, from the first she filled a unique place in my life. I am impatient now, as I was then, of the gross and cynical interpretations that are put upon the love of pupil for teacher. It is but one of the many forms of love (I shall have occasion to speak of others in their place), and it can be a strong and beautiful thing. Here, for the first time, I found someone who recognized my passion for science and sympathized with it. In the first year she taught us botany, and in the second, chemistry. I learned to my chagrin that the Linnaean System was obsolete, and I gained a healthy respect for atoms. When she saw my interests burst the narrow bounds of the hours grudgingly accorded to science, she lent me books on physics and took me to museums. One Christmas she lent me a book on astronomy, and adroitly converted it into a Christmas present with the words 'you can keep it if you like'. Adoration ripened into friendship. For many years we met when we could, and took long country walks together. Forty years later I had the joy of presenting my daughter to her, then of the same age as mine when I first knew her.

When I was about 13 I began to write verse, literally in torrents. I had put words together ever since I knew how to write, but it had been a mere verbal exercise. The verse was something of another order. It filled my school notebooks. I distributed it to my friends. I wrote at home. At last I even showed it to a friend of my Mother, who had actually published a book of poems. She gave me excellent advice: practice writing in strict forms to discipline yourself. I betook myself to the sonnet, the vilanelle, the rondel, the rondeau redoublé. It never attained the rank of poetry, but I wrote much properly-constructed verse. It taught me the art of writing as nothing else could have done. If you can write good verse (by which I do not mean great poetry) you can write good prose.

I was not modest in my undertakings. I projected a didactic poem in (horrors!) rhymed Alexandrines that was to cover all the sciences in 10 cantos, but only the preamble and the conclusion were ever completed. I not only projected, but completed, a tragedy in blank verse on the theme of a scientist who allowed himself to be corrupted by money, and finally atoned by becoming a farmer. I also dedicated a sequence of 30 sonnets to Miss Dalglish, but neither she nor anybody else ever saw them. From this it may be perceived that I had very little sense of humor, and took myself and my scientific passion rather seriously.

Intellectual preoccupations were fortified in another way. At about this time it became evident that I was developing a growth of facial hair. I turned for help to the family doctor, only to be told that there was nothing to be done about it. I burst into tears. 'Never mind, Cecilia,' he said kindly, 'you've got brains. Make something of them.' It was not much comfort, but it stiffened my resolve to become a scientist.

So I sought an outlet in active scientific work. My new-found interest in botany led me to begin to amass a herbarium, collecting and drying the plants that I found on my holiday walks. Miss Dalglish took a dim view of my collection, which she described as 'dried hay'. 'Why,' she asked, 'did I not make a collection of drawings instead?' So I began to keep careful records of my finds, both drawings and dissections. I did not realize at the time what a valuable discipline I was developing.

We were a family of artists. My Mother in the lead, we used to sit in a row on our holiday expeditions, drawing and painting the view. An accomplished painter herself, she had always given us the best in materials. Both my brother and sister had inherited her gifts. He became a fine painter, and his gift for drawing was to adorn all his work when he later became an archaeologist. My sister put hers to use when she became a distinguished architect. I provided something of comic relief when it came to sketching, and was the subject of some hearty teasing for my 'scientific' drawing, but I was developing a power of observation and accurate record, and ended with a portfolio of botanical drawings that was worth far more than the bale of 'dried hay'. At the same time I was acquiring a thorough knowledge of systematic botany, which has added to my pleasure in every part of the world.

Quiet days came to an abrupt end with the outbreak of World War I, when I was 14. It was a time of silent misery for me. The School posted up a 'Roll of Honor', names of relatives who were in the armed forces. My brother was a schoolboy, and the only other male relative I had, my maternal uncle, was fighting on the other side. My Mother herself, though born in Germany, had become British to the core (she never willingly spoke German), but I was deeply conscious that I was not the same as my friends. The scar has never been effaced: it had the effect of making a pacifist of me.

Early in the War I lost my beloved teacher. Miss Dalglish fell ill, and left the teaching profession forever. We corresponded, but it was years before I saw her again. I regret to remember what a hard time I gave to her successors. I suppose teachers of science were difficult to find, and I was not slow to realize how easily they could be confounded.

I continued the uphill fight for a scientific education. The school's strong suit was classical languages, and the powers above decided that I was to be trained as a classical scholar. During those years I was thoroughly drilled in Cicero (and learned to love his terse incisiveness), and went through the whole of the *Aeneid*, which I did not love at all, though the Georgics are still my delight. Then came the study of Greek – Plato, Euripides, Sophocles. Much as I wished I were learning botany and physics and chemistry, I could not be insensitive to this greatest of all literature. Antigone became, and has remained, the greatest of my heroines.

But still I fought for science. When I won a coveted prize at the end of the year I was asked what book I would choose to receive. It was considered proper to select Milton, or Shakespeare, or some writer of similar prestige. I said I wanted a textbook on fungi. I was deaf to all expostulation: that was what I wanted, and in the end I got it, elegantly bound in leather as befitted a literary giant.

I followed up this little victory with another. The first of two nationwide examinations, preparatory to College entrance, covered the usual subjects – mathematics, languages, literature. I asked to be entered also for botany, a subject that I had studied for the most part by myself, with the help of books borrowed from Miss Dalglish. The school demurred, and I had difficulty in persuading them. I remember with a glow of pride that I was placed at the top of the list in botany. By this time I had decided to become a botanist.

The next few years were a confused and unhappy time. I insisted that I must learn advanced mathematics and German (for both would be necessary to a scientist), and no other girl in the school had any such needs or requirements. Most of the pupils were destined for the social world, and several became successful actresses. Few went on to College.

Finally a kindly teacher tutored me in German. I undertook the study of calculus and coordinate geometry by myself, and mathematics took on a kind of mystical significance. At last I was reluctantly turned over to a stern, ascetic teacher of mathematics, who saw in my passion for the subject a passion for herself, which I certainly did not feel. She drove me into a nervous frenzy, and produced a block about the subject that I have never completely overcome. She sneered at me: 'You will never become a scholar.' It was then that the school gave me up. I was told that I must leave, must transfer to another school, that they could do no more for me.

Birth of a dream

It seemed like the end of the world. I was almost 17. My one desire was to go to Cambridge, and to do so I must win a full scholarship, for there was no money with which to send me; and now the school had washed its hands of me. I wanted to be a scientist, and the only subjects I really knew were Latin and Greek. Mathematics had been my hope, and that had failed me. Yet if I had but known it, the powers that decided my fate had done me the greatest possible service.

5

Dramatic interlude

didn't talk

We led a strangely isolated life. There was little social life at home. The few visitors were my Mother's friends, who came to drink tea and talk. No children came to the house. It was not easy to make friends at school. Our day was interrupted by a half-hour interval during which the whole school was herded into the Assembly Hall. One of the girls would play the piano and there might be a little dancing. There was nothing else to do except talk – and what was there to talk about? These intervals were an agony to me, for I was excruciatingly shy. It is true that I excelled in the classroom, and carried off the little triumphs of school without difficulty. When I was still in the lowest class I was placed second in an all-school examination in general knowledge. This was made the occasion of a public castigation of the whole school for having let one of the youngest pupils outdo them. They resented it, and no wonder. I was teased unmercifully, and took refuge in holding my tongue altogether, in class and out of it.

It all seemed dreadfully unfair. The end of the year was marked by two competitions – the essay and reading aloud. I do not recall that we received any instruction or preparation in these subjects. The essay gave me no difficulty; I always carried off the prize, but in reading I was always ignominiously *failed*: it was automatic, for it happened once when I had been absent and had not even taken the examination. And I tried so hard! I know now what the trouble was. I have always been a very fast reader, and my eye outran my tongue. To this day I cannot read aloud. I am impelled to paraphrase, to condense, to improvise.

In this atmosphere it is not surprising that I made few friends. Our

104

family formed and filled its own little circle. In those prehistoric days before radio and television, our world was the world of books. Our library was an endless resource.

Presently we discovered the theater. We had the unforgettable experience of attending a performance of *Hamlet* without ever having read the story. We were bemused; we had discovered Shakespeare, and began to read the plays together and commit them to memory. We haunted the 'Old Vic', where the great dramas were made available, literally for pennies. Week by week we were introduced to new delights – *Othello, King Richard II, Coriolanus, The merchant of Venice, The tempest*. I can see the performances now in my mind's eye, remember the names of the players, details of the scenery. *Peer Gynt* and *She stoops to conquer* expanded our horizons. All this was in the war years and was often played to an *obligato* of anti-aircraft fire.

The 'Old Vic' introduced us to opera, too. Not merely to the gay strains of the *Beggar's opera*, but to Mozart – *Figaro, Don Giovanni, The magic flute*, became our household language. We could converse with one another by humming a tune, thus making reference to the words of the associated aria, a delightfully sophisticated conversational code.

This was not, of course, my first introduction to music. My Father had begun my musical education when I was two weeks old, and after his death my Mother kept his memory alive by talking of the music he loved. Handel, he had said, was the greatest of musicians. When I was about eight years old I was taken to the Albert Hall to hear the *Messiah*. I remember that when the chorus rose to take up the melody of 'O Thou that tellest good tidings to Zion' I was so overcome that I burst into tears. In after years I have found just such an emotional release at the trumpet call in *Fidelio*, the three great Madonnas in the first room of the Uffizi, and the sight of the distant Popocatepetl and Icctacihuatl from Mexico City.

On a red-letter day I was taken, at the age of 13, to the first performance of *Parsifal* ever given outside Bayreuth. Equally memorable was the experience of hearing Busoni play the *Chromatic fantasia and fugue*; his final rendering of the theme was the most colossal edifice of sound that I have ever heard.

To this day I miss the London theatre. No other city that I know has so much to offer. My Mother had long known the great actress Geneviève Ward, now a very old woman, as fascinating on stage as off. We used to visit her often, and she sent us theatre tickets regularly. We attended the great performances of the Diaghileff Ballet – *Petruschka, Firebird, Boutique fantasque, Scheherazade* and many more. Nijinski and Pavlova were before our time, but the dancing of Lopokova, Massine and Cecchetti is still alive before me as are the stage settings of Bakst. Theatre, opera and ballet were the lifeblood of my late childhood.

Within our family circle it was only a short step from witnessing drama to performing it. We chose such parts of Shakespeare as could be created by a cast of three, manufactured stage properties of muslin, tinsel and cardboard, and fell to with a will.

As it happened, we found a ready audience. My dream of becoming a scientist seemed to require, as a natural corollary, that I should become a teacher, and I wanted to begin to gain experience as soon as possible. So when a friend suggested that I should teach a Sunday School class I seized the opportunity. My pupils were slum children, very likely from illiterate homes, and I fear I taught them a good deal more science than scripture. We got a good deal of family fun out of drawing cartoons of each other, and I still have one that portrays me addressing a terrified group of small boys: 'And now that you know all about the photosynthetical effects of convection on carbon dioxide, we will go on . . .' We volunteered to give dramatic performances in the Church Hall, which had a real stage, and the offer was gladly accepted.

I wonder what those children made of our presentation of *Hamlet*, in which I doubled as Gertrude and Horatio, my sister took the part of Hamlet, and my brother those of Claudius, the Ghost and Polonius (we confined ourselves to the first Act and the Closet Scene), or of the even more ambitious production of *King Richard III* (my favorite part), sketchily filled out by the rest of the family, with a friend pressed into the part of Lady Anne. For us, at least, it was a rich experience, and to this day I pass lonely hours by repeating to myself scene after scene of *Hamlet*, which I committed to memory at that time.

Dramatic interlude

Someone finally suggested to us that our audience might enjoy some other kind of performance more than Shakespeare. The entertainment of the children of St Mark's parish had become our hobby by that time, and we did branch out into *Box and Cox* and *Ali Baba and the forty thieves*, which were well received. Finally I atoned for my earlier blank verse tragedy by writing, producing and acting in a roaring farce, which culminated our dramatic adventures.

These histrionic excursions were not without their value. They formed an excellent introduction to the art of public speaking, for they gave me confidence and taught me the value of emphasis and pacing. I can look back on them with pleasure and gratitude.

6

The dream fulfilled

The move to St Paul's Girls' School seemed like a step from medieval to modern times. There were laboratories for biology, physics and chemistry, and teachers who were specialists in their sciences. I was not only permitted, but actually encouraged, to study science. As I look back I see that life began for me when I entered the doors from Brook Green, Hammersmith. It was a time of dynamic happiness. I remember saying to myself: 'I shall never be lonely again; now I can think about science.'

The school ministered to my twin loves, science and music. Here I came under the spell of Gustav Holst, or 'Gussie' as we affectionately called him. Aside from my shadowy Father, and my schoolboy brother, he was the first man I ever knew. He radiated music; the organ in the great hall reverberated to the great *Toccata and fugue* of Bach. Here for the first time I heard *The planets* (then newly composed) and took part in a performance of the *Hymn of Jesus*. He was like a father to us, shy, abrupt and charming. He was quick to learn of my love of music, asked me to play the violin to him, and urged me to become a musician. I played in the orchestra, and learned conducting from him, but my love of science triumphed. It seems odd to think that the only career I was ever encouraged to follow was that of a musician. As a student at Cambridge I trained and conducted a choir that won an award. One of the judges told me that my conducting had been the decisive factor, and that my future lay there. Indeed, the feelings evoked by conducting a choir or orchestra are so powerful as to be overwhelming, but I recoiled instinctively from something that I felt would control *me*; as a scientist I should be in control of my material. Who knows whether I was right?

There was little more than a year in which to prepare for the fateful scholarship examination. Chemistry and physics had to be attacked almost from the beginning; in botany I had a better start. Chemistry seemed to be an exercise in memory and manual dexterity, but physics opened up a new world. At 13 I had been taught by a radiant, dynamic character; at 17 I first came under the influence of a scientist. Ivy Pendlebury did not teach, she elicited. She allowed me to unfold the subject for myself, drawing conclusion from premise, basing premise on observed fact. How different she was from an earlier teacher, of whom I had written

> Out on you, fond instructor, perverter of Nature's laws,
> Explaining cause by effect, confounding effect with cause!
> I could say a thousand things about you, but I will desist,
> For you are a charming woman, but you are not a scientist.

Miss Pendlebury led me through mechanics and rigid dynamics, electricity and magnetism, light and thermodynamics and the rudiments of astronomy. It is hard to believe, now, how much was crammed into that ecstatic time. She told me that she had never had a pupil with my power of sustained application. It was in fact the releasing of years of pent-up, unsatisfied desire. By the end of my schooldays, physics was replacing botany in my affections.

All motion, I had learned, was relative. Suddenly, as I was walking down a London street, I asked myself: 'relative to *what*?'. The solid ground failed beneath my feet. With the familiar leaping of the heart I had my first sense of the Cosmos. When I tried to tell Miss Pendlebury of the experience, she remarked calmly that I should find Relativity very interesting. She was wiser than she knew – it was Relativity that finally impelled me into the path I was to follow. Or was she? Of all the people I have known, she is the only one that I credit with Second Sight. On the fateful day when I left for Cambridge and the scholarship examination, she looked at me with her grave, penetrating brown eyes and said: 'You'll get that scholarship.' And I did. Perhaps she really knew. The previous day she had instructed me in the theory and construction of an air thermometer. I was not in the least surprised when the examiners made the air thermometer the basis of my laboratory test. It had been foreseen.

Beyond all hope and expectation I was able to go to College. I had won the only open scholarship large enough to pay my expenses. I was ready to enter Newnham College, Cambridge.

Before entering Cambridge University one had to pass the so-called Previous Examination, colloquially known as the 'Little-go'. Because of the confused pattern of my last years at school, I had received no special preparation for this test. One had to qualify in elementary mathematics, Latin and Greek; I believe this was the last year in which Greek was required. I noted with satisfaction that the texts prescribed for this language were the Gospel of St Matthew or the *Apology* of Plato. I had studied the latter, and practically knew it by heart. I had taken my copies of the *Apology* and the *Crito* to a bookbinder, with the request that the words *Holy Bible* should be inscribed on the spine of the volume. This request was indignantly rejected by the worthy bookbinder.

In the examination room I opted for Plato. The Examiner looked hard at me, and exclaimed: 'But you're not a Jew!' In that moment I knew for the first time that my world made a distinction between Jew and Gentile. Only later did I realize that the obligatory Greek Gospel bore the mark of the day when the College degree was the gateway to the ministry of the Church of England. The alternative of Plato told the tale of the more recent admission of non-Christians to the University. There was no appeal from the rules. I was a Christian, and St Matthew it must be. I had never studied the Greek text, but thanks to my intensive religious training (which I now blessed for the first time) I knew the English version practically by heart. With the aid of a few key words I was able to identify the passages, and to produce, not a translation, but an accurate repetition.

There had been no anti-Semitism in the world I had known. So many of the girls at St Paul's were Jewish that Kosher luncheons were provided for them. We used to be asked: 'Regular or Special?' and I acquired a taste for the 'Special', from which the school, unlike Cambridge University, did not debar me. It was much the same as the question: 'Fish or meat, Miss?' that had come at lunch every Friday at my previous school. I felt that to choose fish was a display of smugness, and always defiantly selected meat. The irreverent used to parody the Church Catechism: 'My duty towards Cod is to eat the beastly stuff.'

The dream fulfilled

Not until I came to the United States did I encounter real anti-Semitism, and it came as a shock to me. I expressed to a friend that I liked one of the other girls in the House where I lived at Radcliffe College. She was shocked; 'But she's a *Jew*!' was her comment. This frankly puzzled me, for the girl in question made quite a point of being a Christian, which should, I felt, have exonerated her in my friend's eyes.

I found the same attitude towards those of African descent. At that time a coloured skin was such a rarity in England that I do not think I had ever come across one personally. I had known that many Americans were black, and accepted them as a natural part of the population. I well remember the surprised looks in the room when I invited the colored janitor, called in for some routine inquiry, to take a chair.

This is not to say that I grew up free of race prejudice. An English girl with my background inherited a distrust of the Irish and the Asiatic Indian. I drew up a race prejudice map of the world and concluded that it is always produced by economic and social tensions. I do not see how we can escape from these, but we can recognize them, and a prejudice that is frankly recognized is half conquered. It is the unrecognized, blind prejudice that is fatal.

The mandatory Greek section of the 'Little-go' started a train of thought that has led me into a digression. It reminds me of a later experience, for Greek proved to be more than a passport to Cambridge. Ironically, it was the part of my education that I first put to economic use. In need of money at the end of my Cambridge days, I made a shameless search of the College catalogue for a profitable prize. There I found an offer of £50, a veritable fortune, for an essay on the Greek text of one of the Gospels, with special reference to some doctrinal point. I secured a ticket to the Reading Room of the British Museum, made a digest of a dozen theological treatises, and produced a long essay, laden with Greek footnotes. As I had guessed correctly, the subject had attracted no other candidates, and I was able to enjoy my first earnings, which I spent on outfitting myself for my journey to the New World.

7

Pathway to the stars

In September 1919 I entered Newnham College, Cambridge. The atmosphere was euphoric. The 'War to end war' was over. Few among us young people doubted that the Millennium was upon us. As we entered College, we were swept up by the surge of politics. We women, of course, had no votes (even had we been old enough), but that did not prevent us from conducting spirited debates. A new world was opening before mankind. We declared almost unanimously for Labour.

I do not think any of us doubted that war was gone forever. I had stood in a London street and heard the delirious acclamation as Woodrow Wilson rode by, waving to the crowds and smiling a fixed, sardonic smile. The League of Nations was to save the world, and he was its prophet. Humanity was saved, and now we could turn to other things. That, as I see it now, was the mistake of my generation.

The intellectual atmosphere was equally heady. Ernest Rutherford had come to the Cavendish Laboratory from Manchester as Professor of Physics, and the New Physics was gaining momentum.

As a student in natural sciences I had to select three subjects for the first part, and (if one proceeded to the second part, which many did not) a single one for the second. Officially I was still destined for botany, which would normally have been combined with zoology and chemistry, but I insisted on physics and was allowed the unusual combination of botany, physics and chemistry. The lure of the Cavendish Laboratory was irresistible.

From the first I found botany disillusioning. My recent reading had been concentrated on paleobotany. Unable to afford to buy books, I had transcribed Scott's *Fossil botany* into a notebook during my last year at school. I knew that the Professor of Botany, A. C. Seward, was a leading paleobotanist, and hoped great things, but I had to attend the elementary lectures, whose content was familiar and boring. We were not encouraged to go beyond their limits. I found a fascinating group of desmids among the more humdrum algae that we were set to study under the microscope, and I told the demonstrator that I was having difficulty in identifying them. He turned me off relentlessly: 'They don't come into your course.' Poor man, desmids were probably not his subject. But I felt cheated. There were no lectures on Paleobotany and when I actually met the Professor he furnished neither encouragement nor inspiration.

There had been other rebuffs. By now I was an avid collector, fascinated by the ramifications of systematics. I tried to convince my teachers that *Adoxa* belonged with the Saxifragaceae. I found a remarkable rose on the cliffs of Cornwall, and carried a drawing of it to the Herbarium of the Natural History Museum in London, convinced that I had found a new species. The Curator, kindly, peppery old Dr Baker, examined my drawing and remarked that the genus *Rosa* is 'very difficult'. Then he added: 'I suppose you tagged the bush?' I was crushed by the enormity of my inexperience.

At about this time, through the kind offices of Aunt Dora, I was invited to visit the experimental plant breeding station of the great William Bateson. He took great pains to show me the Mendelian experiments of which I had read so much. I ventured to express the opinion that 'it must be wonderful to do research'. He turned on me almost savagely, a great craggy man with thick hair and a sweeping moustache. It was *not* wonderful – it was exasperating. It was as easy to study genetics by looking at the types in any crowd of humans. And then, I suppose to shock me, he threw a glance at Wimbledon Church, which overlooked the gardens, and remarked sardonically: 'We learn a lot of hymns here.' I *was* shocked, not by what he said, but by his manner, and I told my Mother so. 'Poor man!' was her comment, 'Science is his religion.' 'Well, it's mine too' I replied at once, and my Mother was shocked in her turn. But cruel as he was, I owe a debt to William Bateson. He almost reduced me to tears, but he taught me a valuable lesson.

For I had a lot to learn about the nature of research, and I have been learning it gradually all my life. At a very early age (still in the Wendover days, for I remember the locale clearly) I made up my mind to do research, and was seized with panic at the thought that everything might be found out before I was old enough to begin! Much later I spoke of my aspirations to Miss Pendlebury. Why, she asked, did I want to do research? My response was immediate: 'It will be so wonderful to make new theories.' She asked whether I realized how very few people make new *theories*; most scientific work consists of making accurate *observations*. I am not sure that the lesson has completely sunk in even yet. The relation between theory and observation is but another aspect of the ancient problem of Faith and Works.

For a year I persevered with botany, attended lectures, drew anatomized plants under the microscope, wrote essays for my tutor. I was fortunate to be tutored by Agnes Arber, beautiful, scholarly, withdrawn. I never came to terms with her view of the subject. I still find her *opus magnum, The natural philosophy of plant form*, to be beyond my understanding. She wrote of 'the facile Darwinian view – so easy to understand, and therefore so fatally easy to accept'. What had Aristotle, Theophrastus, Goethe, Boethius and Spinoza to do with the panorama of the plant world, so neatly classified, so convincingly understood in terms of Natural Selection? Already I had the makings of a closed mind.

The lectures spread before us an array of intricately classified plant species, products of organic evolution (which I equated with Natural Selection). Mendelian heredity was well entrenched, with Bateson as its prophet; the mutation theory was gaining momentum. The physiology of plants was burgeoning into biochemistry. Hopkins was credited with having discovered 'the chemical basis of species'. The spirit of a new, rational biology was in the air, and the old picture seemed as *démodé* as the Ptolemaic system.

Mrs Arber was patient with me, but she could not curb my passion for speculation, for simplistic ideas, 'so fatally easy to accept'. A crisis came when I submitted to her an essay on the evolution of root structures, of which I had a high opinion. She made the crushing comment that my ideas were neither original nor of much significance. I took the criticism very seriously. If I did not know a good idea from a bad one, I was probably studying the wrong subject. This was, of course, barefaced rationalization. I was eager to be off with the old love and to embrace the new, my beloved physics.

I finally saw the writing on the wall in the shape of the structural formulae of the anthocyanin pigments. The study of plant physiology led to a course in biochemistry. I was not ready for it; my knowledge of organic chemistry was rudimentary, and Willstätter's treatise on the anthocyanin pigments defeated me. I saw before me a science based on an intricate empirical foundation, which seemed to be utterly unreal. Were these structural formulae mere parables? Could we ever know what complex organic molecules were really like? Many years later, when I saw Stanley's electron microscope pictures of the tobacco virus molecule, I looked back ruefully at the day when I had decided to turn from an empirical science to one in which one knew what one was talking about. And of course I was wrong; the outer structure of the atom and of the atomic nucleus are an exact parallel, even more elusive than the structure of the organic molecule. But the process of rationalization ran its course, and I made up my mind to turn to physical science.

The most stimulating lectures that I heard at Cambridge were those on physical chemistry by H. J. H. Fenton. He began at the point where most instructors stopped. He laid the facts before us and outlined the accepted theories. Then, with almost derisive eloquence, he proceeded to demolish the elaborate façade. He pointed out other facts, for which the theories did not account, and other theories that covered the facts equally well. He threw his nets over similar trends in other sciences, pointing out the parallel between the Principle of Le Chatelier, the Principle of Least Action, the Hamiltonian Principle, weaving together chemistry, physics and astronomy. His style of lecturing is the one that I have always wished to emulate, but I am conscious that I lack his breadth of knowledge, his unbridled imagination and his unbiased mind. I never had another teacher who combined these qualities.

The study of physics was pure delight. The Cavendish Laboratory was peopled with legendary figures. The great J. J. Thomson hovered in the background; there were Aston with his mass spectrograph, C. T. R. Wilson with his cloud chamber; there were visiting lecturers such as Niels Bohr and Irving Langmuir; and, looming over all, was the figure of Ernest Rutherford. We first-year students saw little of him, but he was always on the horizon, a towering blond giant with a booming voice.

At first we were soaked in classical physics, lectures and laboratory work. There was Alexander Wood of the golden voice, revered as a Muscular Christian. 'I canna believe,' he declaimed with his Paisley accent, 'that the Univairse is a collosal prractical jooke on the parrt of the Creatorr.' The laboratory work was the province of Dr Searle, an explosive, bearded Nemesis who struck terror into my heart. If one made a blunder one was sent to 'stand in the corner' like a naughty child. He had no patience with the women students. He said they disturbed the magnetic equipment, and more than once I heard him shout 'Go and take off your corsets!' for most girls wore these garments then, and steel was beginning to replace whalebone as a stiffening agent. For all his eccentricities, he gave us excellent training in all types of precise measurement and in the correct handling of data.

In spite of his brusque manner, Dr Searle was a kindly man, as I found to my embarrassment. When I felt that I was not keeping up with the heavy program of work, I appealed to him for help. 'There's nothing wrong with your mind,' he assured me, 'It's your *soul* that needs attention.' And he carried me off there and then to a Christian Science healer. It was a very strange confrontation. I sat through the ordeal in silence and thanked her politely, but I never appealed to Dr Searle for help again.

Physics was at the parting of the ways. The classical branches of the subject were indeed enough to fill the student's time if he pursued them in their lapidary symmetry and elegance through mechanics, electromagnetic theory and thermodynamics, but radioactivity was in the air. It dominated the Cavendish Laboratory, for Rutherford was beginning his attack on the atomic nucleus. The legendary J. J. Thomson, discoverer of the electron, was very much alive. James Chadwick, who was to discover the neutron, was a demonstrator in the advanced laboratory. The Bohr atom was introduced to us by Bohr himself. I still have the notes I took during his lectures on atomic structure, my first introduction to the subject that was to dominate many years of my work. His discourse was rendered almost incomprehensible by his accent; there were endless references to what I recorded as 'soup groups', only later emended to 'sub-groups'. We heard rumors, at first little more, of the Quantum Theory. The word went around that a student could attain a First Class without studying classical physics at all.

The stress was on observation. 'One thoroughgoing experiment,' Rutherford thundered in one of his lectures, 'is worth all the theories in the world – even if those theories are those of a Bohr.' Years later, Eddington uttered the dictum that he would not believe an observation unless it was supported by a good theory. I was an astronomer by that time and knew him well. I told him that I was shocked by his pronouncement. He smiled gently. 'I thought it would be good for Rutherford,' he said.

Such was the scientific panorama as I saw it at the end of 1919. I was standing before the door through which I was soon to enter that world for myself. Suddenly, dramatically, it swung open.

There was to be a lecture in the Great Hall of Trinity College. Professor Eddington was to announce the results of the eclipse expedition that he had led to Brazil in 1918. Four tickets for the lecture had been assigned to students at Newnham College and (almost by accident, for one of my friends was unable to go) a ticket fell to me.

The Great Hall was crowded. The speaker was a slender, dark young man with a trick of looking away from his audience and a manner of complete detachment. He gave an outline of the Theory of Relativity in popular language, as none could do better than he. He described the Lorenz–Fitzgerald contraction, the Michelson–Morley experiment and its consequences. He led up to the shift of the stellar images near the Sun as predicted by Einstein and described his verification of the prediction.

The result was a complete transformation of my world picture. I knew again the thunderclap that had come from the realization that all motion is relative. When I returned to my room I found that I could write down the lecture word for word (as I was to do for another lecture a couple of years later). For three nights, I think, I did not sleep. My world had been so shaken that I experienced something very like a nervous breakdown. The experience was so acute, so personal, that I felt a stir of surprise when I read an account of this same lecture in James Hilton's *Random harvest*.

The upshot was, perhaps, a forgone conclusion. I was done with biology, dedicated to physical science, forever. The next day I confronted the College authorities with the statement that I was going to 'change my shop' and read physics.

It would have been impossible for me to transfer to astronomy, which was, and still is, treated at Cambridge as a branch of mathematics, to be approached by way of the Mathematical Tripos. But I could attend all the lectures on astronomy, and I fell on them with avidity at the beginning of the next year. Meanwhile I applied myself to completing the first part of the Natural Sciences Tripos, which I did in two years instead of three, and attained the desired First Class. The rest of my time at Cambridge was to be devoted to physics, with all the astronomy I could pick up on the side.

The advanced course in physics began with Rutherford's lectures. I was the only woman student who attended them and the regulations required that women should sit by themselves in the front row. There had been a time when a chaperone was necessary but mercifully that day was past. At every lecture Rutherford would gaze at me pointedly, as I sat by myself under his very nose, and would begin in his stentorian voice: '*Ladies* and Gentlemen'. All the boys regularly greeted this witticism with thunderous applause, stamping with their feet in the traditional manner, and at every lecture I wished I could sink into the earth. To this day I instinctively take my place as far back as possible in a lecture room.

The laboratory work of the advanced course was exacting and I think the demonstrators shared Rutherford's scorn for women. At this point I needed a good tutor, but Newnham did not provide one in advanced physics, and I was passed from one reluctant young physicist to another, never getting the advice I needed. I was still agonizingly shy, and quite unaccustomed to dealing with men. I was afraid to ask questions, made many blunders, and learned very little about experimental physics.

Rutherford's daughter Eileen, then a student at Newnham, became one of my friends. She was a lovable, spontaneous girl, without an ounce of science in her makeup. When she invited me to her home for tea, I found her Father quite as alarming in private as in public. I was horrified by the zest with which he told a story of 'the man whose life was saved by Swedish Punch'. During the recent War he had dined too well with friends in Sweden and spent the night in the gutter as a consequence. Meanwhile, the ship to which he belonged had sailed without him and was torpedoed in the North Sea. It did not strike me as an amusing story.

Later Eileen reported to me that her father had remarked: 'She isn't interested in *you*, my dear; she's interested in *me*.' I was outraged, for I was in fact very fond of her. There may have been a grain of truth in his remark, nevertheless. This did not allay my anger. Never, never would I rely on private influence, or enter their house again!

By this time I was reading all the astronomical books I could lay my hands on. Eddington's *Stellar movements and the structure of the Universe* introduced me to stellar motions; I did not know that the 'Universe' of that work was the tiny province that Walter Baade later described as the 'local swimming hole'. I discovered Henri Poincaré. I remember finding *Science and hypothesis* in the library at Newnham, sitting down on the floor and reading it from cover to cover on the spot. This was followed by *Science and method*, *The value of science*, and finally *Hypothesès cosmogiques*, a perennial source of inspiration.

Presently I learned that there was to be a public night at the Observatory. I bicycled up Madingly Road and found the visitors assembled in the Sheepshanks Telescope, that curious instrument which, in the words of William Marshall Smart, 'combined all the disadvantages of a refractor and a reflector'. He himself was there when I arrived and I heard him say: 'Avoid the measuring machine with care' – advice that I did not follow later on! The gruff, kindly Second Assistant, Henry Green, was adjusting the telescope, and presently I had a view of a double star whose components (as he pointed out) differed in color. 'How can that be,' I asked him, 'if they are of the same age?' He was at a loss for an answer and when I persisted in my questions he gave up in despair. 'I will leave you in charge,' he said, and fled down the stairs. By that time he had turned the instrument to the Andromeda Spiral. I began to expatiate on it

(Heaven forgive my presumptuousness!) and was standing with a small girl in my arms, telling her what to look for. I heard a soft chuckle behind me, turned round and found Eddington standing there.

As I heard him tell it later, when I had come to know him, Henry Green had gone to 'The Professor's' study and told him: 'There's a woman out there asking questions,' and asked for help. The moment had come and I wasted no time. I blurted out that I should like to be an astronomer. Was it then or later that he made the reply that was to sustain me through many rebuffs? 'I can see no *insuperable* objection.' I asked him what I should read. He mentioned several books, and I found that I had read them all. So he referred me to the *Monthly Notices* and the *Astrophysical Journal*. They were available in the library of the Observatory which he said I was welcome to use. To paraphrase Herschel's epitaph, he had opened the doors of the heavens to me.

I began to attend Eddington's lectures. Those on Relativity revived the interest first stimulated in the Great Hall of Trinity College. The Determination of Orbits and the Reduction of Observations proved to be of more lasting value. Under his eye we computed the orbits of several comets – all, of course, with the use of logarithms. These computational sessions were topped off by the special treat of tea at the Observatory, at the invitation of old Mrs Eddington and her sweet and gentle daughter Winifred. There were only three or four students at these sessions and we were warmly received in the family atmosphere. It came as a slight shock to me to learn that Eddington's favorite composer was Humperdinck, and that the music he liked best included the songs of Harry Lauder, especially *Roamin' in the gloaming*. He was a very quiet man and a conversation with him was punctuated by long silences. He never replied immediately to a question; he pondered it, and after a long (but not uncomfortable) interval would respond with a complete and rounded answer.

There were other lectures too: Smart on Celestial Mechanics and Lunar Theory (of which I did not get the full import, for he was not one to temper the wind to the shorn lamb). Stratton lectured on astrophysics, introduced us to stars, the Russell diagram (as we called it in those days), variable stars and novae. Stratton was the

Director of the Solar Physics Observatory, and here too I was made welcome. When I spoke to him of my desire to become an astronomer, he was less than encouraging. 'You can never hope,' he said, 'to be anything but an amateur.' And there was old Professor Newall, picturesque with his shock of white hair. He announced a course of lectures on solar physics, but they were not a success; the audience fell off progressively and finally even the lecturer himself failed to appear. His looks were portentous, but his performance was not impressive. I was much diverted to hear that old Mrs Eddington had enquired about her son: 'Is Stanley *really* as clever as Professor Newall?'

One afternoon I bicycled up to the Solar Physics Observatory with a question in my mind. I found a young man, his fair hair tumbling over his eyes, sitting astride the roof of one of the buildings, repairing it. 'I have come to ask,' I shouted up at him, 'why the Stark effect is not observed in stellar spectra.' He climbed down and introduced himself as E. A. Milne, second in command at the Observatory. Later he became a good friend and a great inspiration to me. He did not know the answer to my question, which continued to exercise me, as I shall tell later.

At about this time I discovered that Newnham College had an Observatory, which had long stood unused in the grounds. There was a small visual telescope, and I began to explore the sky for myself. The clock would not run but I set that to rights by removing a chrysalis from the works. I discovered the beauties of the planets – who can ever forget his first sight of the moons of Jupiter and the rings of Saturn? I organized public nights and began to observe variable stars and record their changes, and I installed an observing book, with a notice that anyone who observed with the telescope must make a record in the book, and sign and date the entry. I wonder how many records have been made in it since I left it?

A very real friend appeared in the shape of L. J. Comrie, who was working in Cambridge for an advanced degree. He was an extraordinary man. Crippled and deafened by war wounds, he was still extremely active and played a formidable game of tennis in spite of having lost a leg. He had just performed the remarkable feat of making a correct prediction of the eclipse of one of Saturn's satellites by the shadow of another. He gave me valuable lessons in computing, and helped me to acquire a library of mathematical tables. He persuaded me to join the amateur computing section of the British Astronomical Association, which carried on predictions of stellar occultations. Here I found to my chagrin that I was an inaccurate computer and learned, painfully, to check my calculations – an invaluable training that has stood me in good stead.

I began to keep a notebook in which I listed the problems that I should like to study. My first flight was an attempt to interpret Cepheid variables in terms of the oscillations of a star between an oblate and a prolate spheroid. I did not learn of Plummer's work on this problem until many years later. More important, I noted that the absorption of light in space should be studied by discussing separately the colors of stars of all spectral types. I realized that this problem was then unsolved. I did not know that at the end of my scientific life I should still be concerned with it.

The time had come when learning from others was not enough. I wanted to explore the frontiers for myself. I went to Eddington and asked him to introduce me to research. At the time he was working on stellar interiors, and he gave me the problem of integrating the properties of a model star, starting from initial conditions at the center and working outwards, layer by layer. I fell to with enthusiasm; the problem haunted me day and night. I recall a vivid dream that I was at the center of Betelgeuse, and that, as seen from there, the solution was perfectly plain; but it did not seem so in the light of day. After a time it occurred to me that it would be interesting to take the rotation of the star into account, and I rushed in where angels might have feared to tread. As I should have foreseen, I ran into insoluble problems at the stellar surface. Finally I took my incomplete solution to Eddington and asked him how to overcome the difficulty. He smiled. 'I've been trying to solve that problem for years,' he said.

A second problem proved more tractable. Would I measure the proper motions of the stars in the vicinity of the open cluster NGC 1960? So I spent many happy hours (when I ought to have been in the advanced physics laboratory) sitting at the measuring machine in the housing of the Sheepshanks telescope. Finally the measures were all made, and ready for reduction. Dr Smart, who was advising me, told me to obtain the Equations of Condition by Least Squares. I was too proud to own that I did not know how to solve equations by Least Squares. I repaired to the Reading Room of the British Museum and asked for the works of Gauss. My heart sank when I was presented with five huge volumes, all in German. But somehow I subdued the problem, reduced my observations, and wrote them up. Eddington read my little paper, and wrote me the comment that my results were 'very nice'. I submitted it to the Royal Astronomical Society and had the unspeakable joy of seeing my first paper in print.

Each of these two essays in research taught me a valuable lesson. From the first I had learned my limitations. The second taught me that one should never be ashamed to admit one's ignorance. An admission of ignorance may well be a step to a new discovery. To realize one's limitations marks the awakening of intellectual integrity, without which imagination, ingenuity and assiduity are barren.

Why does a man want to be a scientist? There are many goals: fame, position, a thirst for understanding. The first two can be attained without intellectual integrity; the third cannot. I suppose there are few who can resist the lure of fame. A Henry Cavendish is rare. But many a man is famous on account of his less important contributions to knowledge. I often wonder what work such men would prefer to be known for. Position confers power, but power can be a burden to a dedicated scientist. The thirst for knowledge, what Thomas Huxley called the 'Divine dipsomania', can only be satisfied by complete intellectual integrity. It seems to me the only one of the three goals that continues to reward the pursuer. He presses on, 'knowing that Nature never did betray the heart that loved her'. Here is another kind of love, that has so many faces. Love is neither passion, nor pride, nor pity, nor blind adoration, but it can be any or all of these if they are transfigured by deep and unbiased understanding. In the words of T. H. Huxley: 'Sit down before the facts as a little child, be prepared to give up every preconceived notion, follow humbly wherever and to whatever abysses nature leads, or you shall learn nothing. I have only begun to learn content and peace of mind since I have resolved at all risks to do this.'

My college life drew to a close. Nearly all my friends were looking for teaching positions, for teaching at a girls' school was virtually the only kind of work to which we could look forward. I remembered the time when I had longed to be a teacher, to bring to others the joys that scientific learning had brought to me. But a new light was thrown upon the subject by a little book, entitled (I think) *The compleat schoolmarm*, which had just appeared, and was passed from hand to hand with cynical chuckles. It drew a picture of a narrow feminine society, beset by petty jealousies and bounded by small horizons. I saw an abyss opening before my feet. My taste of the world of scientists had unfitted me for such a calling and suddenly I saw the life of a schoolmistress as 'a fate worse than death'. I have never seen the book since then, but I feel indebted to the author (or authoress?).

There was no future for me in England other than teaching. It was Mr Comrie who came to my rescue. He was soon to go to a position at Swarthmore College in the United States, and he told me that a woman would have a better chance to be an astronomer in that country. He offered to take me to a lecture to be given in London by Harlow Shapley, newly appointed Director of the Harvard College Observatory.

It was a memorable lecture. The name of the speaker was not new to me, for I had read the papers on globular clusters that he had written at Mount Wilson Observatory. But I had not been prepared for his youth or his style. Eloquence it could not be called, for he had none of Eddington's classic polish. He spoke with extraordinary directness, conveyed the reality of the cosmic picture in masterly strokes. Here was a man who walked with the stars and spoke of them as familiar friends. They were brought within reach; one could almost touch them. He even descended to levity, but it was a levity that spoke of intimacy. The next day I found I could write down all that he had said, word for word.

Dr Comrie introduced me to Harlow Shapley. I came immediately to the point. 'I should like,' I said, 'to come and work under you.' He answered that he would be delighted: 'When Miss Cannon retires, you can succeed her.' Knowing him as I did later, I doubt whether he took me seriously, or gave me a second thought. But I took *him* seriously. I bent all my efforts to getting the support I needed, and collected enough money in Fellowships and grants to finance a year in the United States.

Meanwhile I passed my final examinations, not too creditably. But when I remember how I neglected my studies for work at the Observatory, it is surprising that I secured even a Second Class in the Tripos.

In the fall of 1923, I prepared to leave my native land, to sail to the west as more than one of my ancestors had done. The time for dreaming was over. I was about to enter the real world.

PART II

The light of common day

8

England and the United States

'You have come to a foreign country, but it will be some time before you find it out.' These were the words of Professor Edward Titchener of Cornell, whose home I visited on the first Christmas of my stay in the United States; his daughter was my college roommate. It could not have been better put, as he well knew; he had left England as a young man, many years before.

What should I say to a young person who was about to take the westward road across the Atlantic? You are going to a land where there is no spring, where summer comes in a sudden burst after the rigors of an icy winter. You are going to a country where there are no primroses, where the violets have no scent, where you will seek in vain for purple heather and golden gorse. Never again will you listen for the first cuckoo on a May morning, never again see the blackthorn bursting into blossom, or smell the sharp heady perfume of the whitethorn. To have given up these things may seem a small price to have paid for the opportunity to do the work I loved. But April never comes without bringing a pang of nostalgia. Was not the price, after all, too great? 'Oh, to be in England . . .' Browning's words are not mere poetic imagery, they are a cry from the heart.

It might have been easier if I had come from Sweden, which is not so unlike the rocky, piny New England hills. Or if I had been bound for the gentler climate of Virginia, or the lovely coast of Oregon. Sadly I must confess that in Massachusetts I have found a 'stony-hearted stepmother'. It has been my home for more than 50 years, but only gradually have I come to accept and love the pine woods and the black-eyed Susans and the milkweed. Every April brings the wave of nostalgia. Not for 45 years was I able to listen once more to the cuckoo in a Buckinghamshire lane.

Fig. 4. Cecilia Payne on board the *Caronia* 1923

What do we mean when we say 'I love my country'? Love, the recurrent theme of life, has many faces. My love for England is, literally, a love of the country – the fields and hedges, the firm turf underfoot, the scent of wild thyme, 'the ousel-cock so black of hue, with orange-tawny bill, the throstle with his note so true, the wren with little quill'.

But England is my country no longer. When, after several years, I knew that my future lay in the United States, I did not hesitate to become an American citizen. My first reason was the conviction that a responsible member of the community should be a voter. Furthermore, I owed gratitude and loyalty to a society that had given me the opportunity to live the life and follow the profession that I so much desired. The same gratitude and loyalty lie behind a married woman's adoption of her Husband's name. But neither the new nationality nor the new name can sever the close bonds with one's origins. I shall always love England in a sense in which I can never love any other land.

Love of country carries also a very different implication: the support of a political system and a type of culture. Here the problem of loyalties becomes more complex. I came to a country whose cultural patterns were in many ways unlike those I had known, a country with a totally different political system from the one to which I had been brought up, a different attitude and relation to the international scene. The deceptive similarity of language tends to blur one's perception of such contrasts.

But in half a century both my mother country and my adopted land have changed. The England of today is not the England I once knew, culturally, politically or internationally. Only the land is the same. Culture and politics are concerned with people, and I am a part of them. I can truly say: 'I am happy in the country of my adoption, I work in it and for it.' I may ache for the English countryside, but my home is the United States. After more than 50 years I have formed a clear and objective picture of the country that I set out to visit for a year, and have never left. Actions speak louder than words. I stayed because I wanted to stay, despite all the pangs of nostalgia.

I suppose that every country in the world has changed so profoundly during the past 50 years that the present seems to bear little relation to the past from which it sprang – the past that anyone of my age can clearly recall. Glaringly obvious of a country such as Russia, this is not less true of my own. Happy the man who can say without regret: '*Tempora mutantur, et nos mutamur in illis*'; pitiable are the reactionaries who cannot make this claim.

There come back to me the words of a plaintive German song, heard once in childhood, with its haunting tune.

> Wenn ich den Wand'rer frage,
> Wo komm'st du her?
> Von Hause, von Hause,
> Von Hause komm' ich her.

> Wenn ich den Wand'rer frage,
>   Wo geh'st du hin?
> Nach Hause, nach Hause,
> Nach Haus' ist all' mein Sinn.
>
> Wenn ich den Wand'rer frage:
>   Was druckt dich sehr?
> Ich habe, ich habe
> Ja keine Heimat mehr.

But this is the lament of a reactionary. Bud and blossom may be sweet to the memory, but who would exchange them for the ripened fruit?

In the circle from which I came in England, there was a very definite stereotype of 'the American': he was very rich and very poorly educated. Chameleon-like, my thought is colored by it when I am in my native land. When I visited England in the company of an American friend, she remarked that I was 'ashamed to be known for an American'. It was true: I imagined that everyone would tag me as 'rich and poorly educated'.

In vain I pointed out to my family that all Americans are not rich: there are wealthy Americans, but there are also wealthy Englishmen. In the Victorian atmosphere of my childhood one was slightly apologetic about a man who had a job. 'Of course, he doesn't have to work,' though the socially correct type of work was meritorious. A woman who had to work was frankly pitied: if she 'had no money' she could never hope to marry. Like a breath of mountain air came the sense that in the United States the question was not: 'Does he *have* to work?' but 'What kind of work does he do?'

In all the households I had known there had been servants. My wealthy aunts had dozens; even we, who were frankly hard up, had two or three in my childhood, as well as a gardener and a gardener's boy. And the servants were a perennial source of anxiety: something might 'upset' them, and they would be off, and there would be a frantic search for others. Someone without servants was a 'working woman', a social inferior. In the new country there was no trouble with servants. One could cook and sew to one's heart's content, uncowed by servants and dressmakers. When I came to the United

States I knew that here was a land where those horrible words 'our betters' had no legitimate context. It was an inspiring revelation, though as my experience deepened I learned that I had been overoptimistic. It supplied one of the strongest reasons for wishing to make this country my home, and in due course to become a citizen of it.

Another obstacle to understanding was that the outward signs of wealth were not the same in the two countries. In 1930 I bought a small car and had to do battle with my family's idea that the possession of a car was not only a sign of wealth, but was also ostentatious and therefore slightly vulgar. The automobile became an essential ingredient of life in the United States long before it did so in England. At the time when I was earning a very small salary I paid for my first car with the proceeds of a course of lectures I gave at Wellesley College, lectures that would have been physically impossible without that means of transport. In England this was seen as evidence of great wealth, a reputation that I was too proud to demolish by revealing the facts and figures. In those days I was so hard up that I lived on a diet of potatoes and sausages, because I found them so distasteful that I could eat only a little of them. At this time, too, I bought a piano because I could not live without music (again at the cost of a restricted diet); this too was regarded as a sign of wealth. After all, America was the land of millionaires!

The notion that Americans are poorly educated is even more difficult to scotch. To the superficial and snobbish this means that they do not know how to pronounce Cholmondley, Marjoribanks, St John and Leveson-Gower. Even supposing that they do not, why should they? This attitude is the one that Bernard Shaw makes Caesar ascribe to the Ancient Briton: 'He thinks that the customs of his tribe are the Laws of Nature.'

The difference is more profound than accidents of pronunciation, or even of vocabulary, which is very difficult to assimilate. Often even now the right word eludes me. In moments of weariness, I revert to type, and say 'dustman', 'rubbish-bin' and 'slop-basin'. When I had been in the United States for 20 years I gave a talk before an audience that included William Lyon Phelps. He came up to me afterwards. 'You're English,' he said. 'You used nine words in that talk that no

American would have used.' I expect I still do, but it is not for want of trying.

The real distinction is that Americans are not worse educated, but differently educated. I went to a private school, and my schoolmates were my social equals. We were taught, taught, taught, we were crammed with information, so that a well-educated English schoolchild became a walking encyclopedia. He could produce all sorts of facts, he knew several languages, he could recognize and use an array of selected quotations – English, Latin, Greek. I do not regret this educational background; I am glad that I possess it. The American schoolchild could profit from some of it. But he receives something that we never had. He is early encouraged to develop his own ideas, to become an individual in his own way. Most important of all, his schooling is not restricted to one social class. He has a motley variety of schoolmates from all walks of life.

My education was a specialized preparation for entry into one stratum of society, the one into which I had been born. I went always to a 'private school', and, after my twelfth year, to a school for girls only. Much better for me if I had gone to a coeducational school, in those days something almost unknown in 'middle and upper-class society'. The result was that, save for my own brother, I had scarcely spoken to man or boy until I went to Cambridge, and even then the opportunities for mixing with the opposite sex were limited and carefully overseen. Those who went to the coeducational National Schools were our social inferiors.

By contrast, I found that the schoolchild in the United States usually went to a public school, where education is not geared to a particular class of society, and where a girl meets and studies with boys as a matter of course. The result is a lack of the finely polished specialization that was my lot, and the educational process is forced to take longer if the fine polish is to be attained. For it can be attained by those who pursue it, and when they reach it they commonly possess a. spontaneity and flexibility that is rarely shown by the products of an education such as I received.

I speak of the far past. Things have changed now, both in England and the United States. I hope the international prejudices have changed too, and that the stereotypes of 'Englishman' and 'American' have been obliterated. Air transport and television are great levellers.

Fig. 5. Cecilia Payne behind Everett House 1924

The American idea of the English, I found, was as wide of the mark as the English idea of the Americans. First came the jocular references to the 'embattled farmers' who had 'beaten the redcoats', which implied in me an attitude about as sensible as if I still resented the Norman Conquest. I countered with the mild remark that it seemed to me that the American Revolution had been a Good Thing. Then somebody asked me whether I 'missed the hunting?'. I disclaimed any interest in blood sports; I had never sat on a horse. Finally I was cut down to size by a lady who heard I had been educated at Cambridge and remarked: 'That's at Oxford, isn't it?'

As a graduate student at Radcliffe College I was assigned a room, and a roommate, at a graduate dormitory. The atmosphere was very different from that of Newnham, and I found the freedom novel and intoxicating. At Cambridge we had had rooms of our own. Permission had been required for leaving College after the evening meal. Lights had to be turned off at 11; in scholastic emergencies we were allowed to work later – by candlelight. Now I came and went as I pleased, could work at the Observatory all night if I so desired.

The climate of the New England fall was so stimulating that I found I could work prodigiously long hours. In Cambridge, by contrast, I had always been tired. The damp chill of the fenland air seemed to penetrate my bones, and I think I ached physically all the time I was there. In the heady atmosphere of New England in October, nothing seemed impossible. Once I worked for 72 hours without sleep. It was indeed a new world, a new life.

My fellow students were kind and friendly, and at first they found me hilarious. They ran around in all stages of dishabille. I do not think that I had been undressed before anybody since I was a baby, and I suppose they found me ridiculously prudish. When they found that I wore layer upon layer of underwear, they used to watch me disrobe with shrieks of incredulous delight. It was not long before their laughter, and the climate, weaned me from my native wardrobe. To dress like the other girls conveyed yet another sense of freedom.

Language differences were gradually revealed to me. At a family dinner to which I had been invited, the daughter of the house used the word 'bloody', too profane to be heard from the lips of an English lady. Years later I met her again, and she recalled how she had done it to shock me. 'You turned very white,' she recalled, 'but you said nothing at all.' I am glad, at least, that I did not blush.

9
Harvard College Observatory

When I arrived in Cambridge, Massachusetts, I had crossed a gulf wider than the Atlantic Ocean. I had left the world of dreams and stepped into reality. Abstract study was a thing of the past; now I was moving among the stars. I had not been wrong in my estimate of Harlow Shapley. His mind travelled about the stellar universe as in familiar country. The Observatory of which he was the newly appointed Director was a storehouse, crammed with the raw material of the Cosmos. Let me give a sketch of the new world in which I found myself.

The day after I arrived in the United States I was installed at a desk in the 'Brick Building', then the newest structure in the Observatory grounds, now the only survivor of the buildings I first knew. It had been built to house the great collection of photographic plates that had been amassed by the previous Director, Edward C. Pickering. Here also were the offices of those who worked with the plates.

In Pickering's days all this work had been done by women. It was respectable work at the end of the nineteenth century, work that conferred a certain distinction, and the old Director had taken full advantage of the fact. The college-trained scientists Henrietta Leavitt, Annie Jump Cannon and Antonia Maury were at the top of the hierarchy. The work of examining the plates, identifying the stars on them, of recording, of proofreading, of computing, of keeping half a million photographs in order in the plate stacks, employed a large team of women, nearly all of whom dated back to the Pickering era when I first came. Someone quoted to me an ancient joke: 'Why is the Brick Building like Heaven? Because there is neither marrying nor giving in marriage there.' Indeed in Pickering's day there had

Fig. 6. Harvia Wilson, Adelaide Ames, Cecilia Payne on Observatory grounds 1924

been no men there. Professor Bailey had an office under the dome of the 15-inch telescope (the 'Great Telescope', in its day the largest in the world), Leon Campbell was even farther away, and Professor King and Professor Gerrish occupied another building on the far side of the grounds. But in 1922 the new Director had broken with tradition, and the young Dutchman Willem Luyten had an office on the lower floor of the plate stacks.

The Brick Building as I first knew it must have retained much of the flavor of Pickering's day. Miss Cannon reigned supreme, she was

'Curator of the Astronomical Photographs'. Much has been written about Miss Cannon; she was a legend in her own day. In the next chapter I shall say something about her work. When I think of her as a person I am at a loss for words to convey her vitality and charm. She had lost her hearing in youth, but she had none of the suspicious pessimism so often associated with the deaf. She wore her hearing aid with an air, and made a virtue of necessity by unshipping it when she wanted to be undisturbed or to do concentrated work. She was warm, cheerful, enthusiastic, hospitable. Like many people who are hard of hearing she had a sharp metallic voice, and often broke into a characteristic, resounding laugh.

Miss Cannon was extraordinarily kind to me. She might well have resented a young and inexperienced student who was presumptuous enough to attempt to interpret the spectra that had been her own preserve for many years. She never gave a sign of doing so. 'Do you realize,' Shapley asked me later, 'how easily Miss Cannon could throw a monkey-wrench into the works for you?' With the arrogance of youth I had not thought of it; I had even permitted myself to wonder how anyone who had worked with stellar spectra for so long could have refrained from drawing any conclusions from them. She was a pure observer, she did not attempt to interpret. As I look back I see how her work has outlasted my early efforts at interpretation. The *Henry Draper Catalogue* is a permanent monument.

I remember a characteristic gesture. Years later, when she was quite old, we gave a spring party in our country garden. The rain came down in torrents and the spirit of the party was greatly damped. But not Miss Cannon's. She arrived in a gay, flowered dress. 'I thought this would do something to counteract the weather,' she laughed. She was never daunted; her spirit was always gay. She died in her seventies, working to the last. But she was one of those whom the gods love, for she died young.

Miss Maury was not a regular member of the group when I first arrived, but she was a frequent visitor. There was a profound difference in scientific outlook between her and Miss Cannon. Miss Maury was a dreamer and a poet, always vehemently denouncing injustice, forever battling for a good (often a lost) cause. She had a passion for understanding things. It was typical of her that she

Fig. 7. Front row (from left): Agnes M. Hoovens, Mary B. Howe, Harvia H. Wilson, Margaret Walton Mayall, Antonia C. Maury
Middle row: Lillian L. Hodgdon, Annie J. Cannon, Evelyn F. Leland, Ida E. Woods, Mabel A. Gill, Florence Cushman
Back row: Margaret Harwood, Cecilia Payne, Arville D. Walker, Edith F. Gill at Observatory 1925 (Harvard College Observatory photograph)

devoted years to the mysteries of Beta Lyrae and Upsilon Sagittarii, still incompletely solved. On Beta Lyrae she liked to quote Miss Leavitt: 'We shall never understand it until we find a way to send up a net and *fetch the thing down*!'

Like Miss Cannon, Miss Maury was extremely good to me. Many were the long talks that we had about the problems of stellar spectra. We both liked to work at night, and our discussions were painfully punctuated by insect bites, for she insisted on keeping the windows open and could not bear to kill the mosquitoes.

I have often speculated on the impetus that is given to a person by some handicap of nature. Miss Cannon and Miss Leavitt were both deaf. Miss Maury's handicap was extreme homeliness, mitigated by her beautiful brown eyes. She did not, indeed, pay much attention to her appearance. She would come to work, her associates said, in one black stocking and one brown. In this she was a contrast to Miss Cannon, who was always smartly dressed, and even in old age was very handsome. Her pictures show that in youth she was extremely beautiful.

The Brick Building was a hive of industry. The *Henry Draper Catalogue* was going through the press, and Miss Florence Cushman had the task of reading the proofs. This dignified galleon of a woman, with her masses of white hair, proudly remembered that Charlotte Cushman, the celebrated actress, had been her aunt. She herself, she told me, had played the part of the fairy Fleta in *Iolanthe* in days long past. It was hard to picture. Miss Cushman was a Lady of the Old School. She told me that she had 'never ridden in one of those new-fangled automobiles'. When I bought a car I persuaded her to take her first ride, and I think she enjoyed it, though she was a little apprehensive too.

Then there was Miss Woods, a stickler for protocol. I think she had acted as Pickering's secretary, and it had left her with a sense of being slightly superior to the rest. '*We* do not use the Harvard Headings,' she admonished me when she found me writing a letter on Observatory notepaper. One of her duties was to examine the photographs for variable stars, and she proudly displayed several medals that she had received for discovering novae – a practice now discontinued. But I never got to know Miss Woods: she kept her distance.

Miss Wells was different. She was tiny and vehement, and she seemed to me very, very old. Years before she had discovered SS Cygni on Harvard plates, and it had been known as 'the Wells variable'. Today SS Cygni is such a classic star that it is hard to believe that I knew the person who discovered it. I have a memory (or is it just a legend?) of her sitting at her desk marking stars on a plate, and then falling asleep and rubbing off all the marks with her nose.

The Gill sisters, Mabel and Edith, hover in the background, quiet, self-effacing, always busy. Miss Hodgden, a voluble strutting hen, has left a clearer mark. Her business was to renew the envelopes in which the plates were kept as they became tattered with use, and she was like the fly on the coachwheel. In her own mind she kept the whole enterprise in motion. Later on she was assigned to do calculations for me. I am afraid she was not a very good computer.

If the spirit of Pickering still stalked the plate stacks, a very different spirit was abroad in the Observatory. The young Director, Harlow Shapley, had begun to transform it. His first step had been the infusion of newer young blood into the astronomical storehouse. In choosing a secretary he had passed over Miss Woods, and taken one of the younger women, Arville Walker. It was an excellent choice. Miss Walker, ('Billy' to her friends) was wise, friendly and discreet. She knew, none better, how to do a good job with accuracy and dispatch. A devout Christian Scientist, she never complained, and was always cheerful and willing. Even when she was struck by a car and gravely bruised, she kept up her work without a murmur. I turned to others for stimulation, but I always went to Billy Walker for advice.

My two contemporaries were Willem Luyten and Adelaide Ames. Willem had come the year before – Shapley's first scientific appointment. He was always ready to argue, always critical and nearly always right. I soon developed a healthy respect for his keen mind, a respect I have never lost. Adelaide was Dr Shapley's scientific assistant, newly graduated from Vassar. She was young, lovely, intensely vital. Though we differed greatly in temperament, she was the closest friend I ever made at the Observatory. She died young, 'to the unspeakable grief of those who knew and loved her'.

On the far side of the grounds were the business office, the domain of Professor Willard Peabody Gerrish, and the photographic department, the province of Professor Edward Skinner King. Although both had the title of Professor, they did no teaching. Mr Gerrish was Professor of Engineering, and the instrumentation was his province. Miss Maury attributed the fine performance of the Harvard instruments to his skill. He had a passion for detail, sometimes carried to extremes. When he retired a box of tiny electric bulbs was found in his office, labelled 'flashlight bulbs, probably burned out'.

Professor King managed the photographic observing, which at that time was carried out in Cambridge. He taught me to use the telescopes. As in everything he did, his watchword was accuracy. He admonished me that one should never record the time of ending an exposure until the shutter had actually been closed; one might die in the interval, and the record would then be inaccurate. He was sensitive and fussy, but we soon found a bond in our love of old books, and he used to bring his treasures, rather furtively, to my office to be admired.

The whole operation was dignified and decorous, but a touch of comedy lurked in the background in the antic, angular shape of Frank Bowie, the night assistant. He moved in other circles. His associates were the underworld characters who managed the Numbers Game. Between taking plates he confided some of the secrets to me. The Right Ascension and Declination of a newly-discovered comet was considered a 'lucky number' and was duly played. On one occasion it paid off handsomely, and enhanced the faith of the underworld in the power of the stars. I can still hear Frank Bowie's raucous voice uplifted in the strains of *Charmaine* as he plied the developer in the darkroom.

After a few years I bought a car – a daring step, for I had never learned to drive. Frank Bowie boasted that he had stolen the first car in Cambridge for a joyride, who knows how many years before? He offered to teach me. Many a time, when dawn put an end to observing, he and I burned up the road beside the Charles River. I can still hear his voice at my elbow, urging me on with 'Step on it, Celia!'

10

The cradle of astrophysics

I make no apology for taking time to look back on the history of the great repository of data that was Edward C. Pickering's legacy. Hitherto I have written of my own small struggles. Now my theme has broadened, and I am concerned with the birth and development of a new science, a drama in which I was privileged to play a modest supporting role.

It is not to deny the contribution of earlier Directors of the Harvard College Observatory to say that Pickering gave a crucial impetus to twentieth-century astronomy. He saw the importance of observations, and set himself to supply those that were crucial to the astronomy of his day. Facts may seem like dry bones to the soaring imagination of the theorist. They are bones indeed, for they constitute the skeletal framework of the science, without which she could neither stand, nor walk, nor take the great leaps that have marked her progress in the last half century. Theory without facts is bombinating in a vacuum. The supposed conflict between the two contains the old mystery of the interdependence of Faith and Works, Body and Spirit.

During the nineteenth century the emphasis of stellar astronomy was on position and motion. It was the era of parallax and proper motion, of double stars and of the great *Durchmusterungen* which located and catalogued the brighter stars in both hemispheres. In the twentieth century the emphasis shifted to the star as an individual.

Pickering's work placed stellar photometry, stellar spectra, and the study of variable stars on a completely new footing. When we look at what he projected and what he achieved, we cannot but marvel at his foresight and his performance. It is said that when he heard of the plans for the *Carte du Ciel,* he decided to carry out the same task of positional recording single-handed. To this end the 24-inch Bruce doublet was fabricated, giving a scale of a minute of arc to a millimeter (the same as that of the Astrographic instruments). The whole sky was systematically photographed, first from Cambridge, then from Peru. The Bruce plates covered the whole sky (except for a few tiny areas), and the fruits of that survey have long been history, as witnessed by the proper-motion studies carried out by Willem Luyten on this great collection of photographs. When compared with the long agony that has attended the completion of the *Carte du Ciel* for nearly a century, the Bruce survey was a marvel of realistic planning. Yet this is but a small facet of the magnificent jewel that was Pickering's legacy.

Accurate photometry is basic to every province of astronomy. The growth of our knowledge of stars and stellar systems during the present century has gone hand in hand with the progress of photometry. The mere visual estimates of the *Durchmusterungen* were replaced by the more accurate measures of the polarizing photometer, and Pickering undertook the enormous task that is embodied in the *Harvard Photometry*. He did not confine himself to planning; he carried out much of the actual observing himself. There was a celebration on the occasion when he made his millionth observation. The *Harvard Revised Photometry* of bright stars was the direct ancestor of the standard *Yale Catalogue of Bright Stars* of today. And the photometry of thousands of fainter stars was also part of the plan; for many such stars it is still the only source of measured information.

The astronomical use of the photographic plate opened up new fields, and here again Harvard planning took the lead. First in importance was the establishment of fundamental photographic standards: the North Polar Sequence and a group of secondary sequences in the Harvard Standard Regions (spaced at intervals of 15° in declination over the whole sky). The techniques were in the hands of E. S. King, the execution in those of Henrietta Leavitt. This was a piece of masterly planning, but it ruthlessly relegated Miss Leavitt to the drudgery of fundamental photometry when her real interest lay in the variable stars that she had begun to discover in the Magellanic Clouds. She was the ablest of all the workers at Harvard at the turn of the century, but Pickering was a dictator, and his word was law. He realized, as Shapley realized after him, that the study of variable stars can lead to precise and significant results only when it is based on accurate photometry.

I need hardly emphasize the difficulties of photographic photometry. I heard it said when I came to Harvard that what really killed Miss Leavitt was Pickering's requirement that she devise a method by which the photographic magnitudes determined with all the Harvard instruments could be reduced to the same photometric system. I cannot believe that he made so unrealistic a request. By the turn of the century the Harvard telescopes ranged through the 24-inch Bruce doublet, the 24-inch reflector, the 16-inch Metcalf doublet, the 10-inch Metcalf triplet, down to the older 8-inch Bache and the one-inch Voigtländer. In the 1920s I myself attempted to reduce their magnitude systems to some sort of order, and was defeated by the variety of their color curves and the consequent intractability of their color equations. It is incredible that Pickering was unaware of these problems, or believed it possible to circumvent them. Miss Leavitt must have encountered the full effect of these difficulties in her classic work on the North Polar Sequence, which coordinated the results from large and small instruments, and called for combination with the results that Frederick Seares was then obtaining for the fainter members of the sequence with the reflecting telescopes at Mount Wilson Observatory.

It may have been a wise decision to assign the problems of photographic photometry to Miss Leavitt, the ablest of the many women who have played their part in the work of Harvard College Observatory. But it was also a harsh decision, which condemned a brilliant scientist to uncongenial work, and probably set back the study of variable stars for several decades. I speak feelingly. When first I went to Harvard, Dr Shapley wanted me to take up Miss Leavitt's unfinished work in photographic photometry, but I had other desires; stellar spectra were my early love, and so they have remained. But I have come to see that standard photometry is the point upon which the pyramid of stellar astronomy is balanced. In the end I reluctantly embarked on the subject, but it was already too late; the emphasis was shifting to photoelectric methods. I shall have something to say later on the growth of photometry. Pickering was wise in his day to invest his finest workers, and the time and resources of his Observatory, in laying the foundations of photographic photometry. Thirty years later it would be superseded.

The Harvard work on the classification of stellar spectra was another marvel of planning and execution. From the beginning Pickering made an unerring choice of aims, means and human tools. Secchi, Vogel, Lockyer and Huggins had been the pioneers, had shown what riches lay in store and had taken the first steps towards reducing them to order. Secchi's observations were visual and lacked detail; his classification was too simple. Lockyer, confining himself to bright stars, revealed so great a richness that his classification, with its Alnitamian and Aldebaranian types, was unwieldy and impractical. In fact he was already taking the leap from classification to analysis, anticipating twentieth-century astrophysics before the world of astronomy was ready to assimilate it.

The Harvard plans called for a massive collection of data, information about as many stars as possible – which meant reaching stars as faint as possible. Hence the employment of the objective prism, which not only reached faint stars, but also recorded many with one exposure. The next step was the systematic arrangement of the information, which soon led to the discovery that the spectra could be placed in an orderly series. Here again, Pickering chose his human agents with skill. Mrs Williamina Fleming was promoted to astronomy from domestic work when her flair for recognizing spectra was discovered. She laid the groundwork of the Henry Draper classification and organized a systematic application of it. She must have been an excellent organizer, a strenuous worker, a martinet. When I first went to Harvard, several years after her death, she was still spoken of with respect, with a kind of awe; but to say that she had not been beloved was an understatement.

The first classification was, characteristically, independent of the previous attempts that had been made to place stellar spectra in an ordered sequence. They were arranged in order, and distinguishable steps were named A, B, C, D, ... without implication or interpretation. Later work, of course, altered the order slightly and led to the dropping of some of the classes, but the major discovery had been made. The spectra of most stars could be arranged in a continuous sequence.

The brightest stars, recorded in more detail by means of objective prisms of wider angle, were next to be classified, and, if possible, analyzed. The work was placed in the hands of the two women who were to be the great contributors to the birth of astrophysics, Annie Cannon and Antonia Maury. Miss Cannon was to study the southern stars, Miss Maury those in the northern hemisphere. The result was one of the classics of astrophysics – Volume 28 of the *Annals of Harvard College Observatory*. It still repays analysis.

Miss Cannon completed her work swiftly. It was typical of her to do so. She placed the Henry Draper system of classification on a firm and definite footing. The one-dimensional scheme remained the standard for decades. Little escaped her keen eye, and her descriptions of the spectra of the brighter stars are still worth reading. Miss Maury, on the other hand, worked slowly, and the section of the catalogue devoted to the northern stars appeared many years after its southern counterpart. She did not employ the letters that were used to define the Henry Draper classes. Her elaborate system was expressed in Roman numerals, and it was two-dimensional: each class had three subdivisions based on the quality of the spectrum lines. Essentially her system anticipated the basic concept of the modern classification developed by W. W. Morgan and his associates, which specifies luminosity classes. For, as Ejnar Hertzsprung was not slow to point out, her third subdivision segregated the stars of high luminosity. But Miss Maury's classification came too late: the Henry Draper system was already entrenched and her classes, though physically more significant, seemed unnecessarily elaborate.

The difference between the two contributions was a reflection of scientific outlook. Miss Cannon had a genius for classification; she was content to leave the work of interpretation to others and, to do her justice, was always ready to greet their findings with delight. Miss Maury had a passion for understanding, a genius for analysis that led her to anticipate by several decades the discovery of luminosity criteria. Her eye for detail was as keen as Miss Cannon's, but she was always slowing things up by asking what it meant.

Pickering's objective was the wholesale classification of stellar spectra, a project that resulted in the nine volumes of the *Henry Draper Catalogue*. He made an unerring practical choice, and placed the work in the hands of Miss Cannon. Pickering chose his staff to work, not to think. Miss Maury was in continual conflict with him. She once made a remark to me that spoke volumes: 'I always wanted to learn the calculus, but Professor Pickering did not wish it.' In her I found a kindred spirit. Once she told me that I was the daughter that she had always dreamed of having.

The *Henry Draper Catalogue* of stellar spectra is a model of wise and realistic planning. The physical means – the objective prism – had been put to work. The observing program was planned and executed to cover the whole northern and southern sky. The system of classification had been defined. Miss Cannon, with her keen eye and extraordinary visual memory, was there to execute the plan. She completed the work and classified a quarter of a million spectra (many of them more than once) in about five years. The resulting catalogue was a model of conciseness, consistency and accuracy. Fifty years after its publication it is still a standard reference work. As a great astronomical project, well planned, well executed, swiftly completed, it is comparable to the *Bonn Durchmusterung*.

I saw the completion of the final stages of the *Catalogue*; the last volumes were going through the press when I came to Harvard. From watching the meticulous proofreading and checking I learned my first lesson in the publication of scientific data. Even more impressive was the sight of Miss Cannon at work, for she had already embarked on the *Henry Draper Extension*. She would take an objective prism plate and would swiftly number the images with a pen. Then, with a recorder at her side, she would classify the spectra, speaking as fast as the recorder could write. When the plate was finished with, it was handed to an assistant who laid it directly on the *Durchmusterung* chart and wrote down the DM numbers of the classified stars. The telescopes with which the plates had been made were so designed that the scale of the plates was the same as that of the charts – one of the many efficiency devices that made possible the swift execution of the program.

When I first looked at the plates from which the Henry Draper classifications had been made I was amazed. The spectra looked like tiny parallel smears, not like the beautiful pictures in Sir William Huggins' atlas, or even the spectra from which Miss Cannon and Miss Maury had made Volume 28 of the *Harvard Annals*. It seemed impossible that anyone could see enough in those tiny smears to classify the spectra. Sometimes, indeed, I would find one of Miss Cannon's numbers in a spot where I could see nothing but a faint blur. Several times I brought her a plate that she had classified years before and asked her to verify a spectral class. Not once did her re-examination produce anything different from her original estimate.

A legend was current that Miss Cannon could remember everything she had ever classified, and could immediately recall the serial number of the plate on which she had examined a particular star. But it is hard to believe that she could remember exactly *how* she had classified each of a quarter of a million spectra.

In the last years of Miss Cannon's life, Henry Norris Russell used to say: 'Somebody ought to find out from Miss Cannon exactly how she classifies each spectral type.' I argued with him that she would not be able to tell them because *she did not know*. She was like a person with a phenomenal memory for faces. She had amazing visual recall, but it was not based on reasoning. She did not think about the spectra as she classified them – she simply recognized them. This was, in fact, one of the strengths of the *Catalogue*.

But sometimes, especially for very faint stars, she did not rely on the line spectrum, but on the apparent distribution of energy in the continuum. This reliance on the continuum was brought out when she classified a number of spectra of faint stars sent from Mount Wilson Observatory. Her classifications were systematically later than from objective prism spectra of the same stars, on account of the relative weakness of the slit spectra in the ultraviolet. When more modern classifications of the spectra of blue stars are compared with hers, the fainter ones tend to be of earlier spectral class than those she assigned. As such stars tend to be reddened, this is no doubt a consequence of her reliance on the apparent distribution of energy rather than (or as well as) on the line spectrum. At this time (as I shall mention later) Shapley did not admit the importance of interstellar reddening, and reliance was placed on the continuum. But apart from systematic errors for faint blue stars, the *Henry Draper Catalogue* is a model of consistency.

The cradle of astrophysics

Thousands of stars were found to be essentially similar to one another. All the more important was the recognition of a few that did not conform. The remarks to the *Henry Draper Catalogue* mention such nonconformists – stars with narrow lines, with abnormally strong lines of strontium and silicon. Those remarks, a testimony to Miss Cannon's keen eye, are still worth reading; they have formed the basis of more than one doctoral thesis.

A third legacy that Pickering left to future astronomers was the study of variable stars, and it was less fully realized than the other two. Stellar spectra could stand on their own feet, though the imperative need for spectrophotometry had not yet declared itself. The mastery of visual photometry had been followed by an attack on photographic photometry, whose problems had been only partially and provisionally solved, despite the North Polar Sequence and the Harvard Standard Regions. Other techniques, photoelectric and electronic, would have to be mastered. The problem of standard photometry remains one of the most difficult in observational astronomy. If those who work in other fields doubt this, let them ask those who have devoted their lives to it. Photometry has occupied, and absorbed, the ablest and the most ingenious: Frederick Seares and Henrietta Leavitt, Karl Schwarzschild, Harold Johnson, Joel Stebbins, William Baum and Gerald Kron, to name but a few.

Photography had given an impulse to the study of variable stars. Mrs Fleming accumulated observations of Mira stars; Miss Cannon's early work was concerned with variable stars. Miss Leavitt found thousands of variables in the Magellanic Clouds, and discovered the period–luminosity relation, probably her greatest contribution. But she was diverted into standard photometry; Mrs Fleming and Miss Cannon became associated with spectral classification.

Another study of variable stars, which made history at the turn of the century, has yet to be mentioned. Solon Bailey discovered large numbers of variables in several globular clusters, thereby laying the foundation of the branch of astronomy by which Shapley first made his name, a subject that has yet to be fully exploited. The astronomer of today discusses variable stars and determines their periods with the aid of the electronic computer. I can look back on earlier days when such work was carried out with the electric desk calculator, and even earlier with the hand-cranked machine. My own earliest work was done with the elephantine Comptometer. But Bailey did all his work on the periods of the RR Lyrae stars with seven-figure logarithms! If I had not seen his working sheets I should not have believed it possible. Most students of variable stars are familiar with his well-known diagrams of the light curves of types a, b, and c. I do not think many of them realize the extraordinary feat that is represented by Bailey's volume in which he published his results on the variable stars in Omega Centauri and Messier 3. It ranks as a classic.

If the Harvard work on variable stars as Pickering left it was a beginning rather than a completed project like the *Henry Draper Catalogue*, it was because the potentialities of the subject were not realized at the time. When the discovery of variable stars began, and a system of naming them by letters within each constellation was chosen, no one guessed how many stars would require names. Perhaps the letters R to Z were originally thought to suffice; then the somewhat clumsy RR, RS, ... seemed to allow for adequate names. Who in those days could have foreseen V2000 Sagittarii? Miss Leavitt, perhaps, with her experience of the Magellanic Clouds. Variable stars were indeed discovered in large numbers on the Harvard plates, but it was left for Harlow Shapley to plan a systematic attack on the problem. Even he, I believe, did not realize the enormous scope of the subject, which I shall discuss in a later chapter.

When I first became an astronomer, variable stars were regarded as a sort of second-class problem. In the last 50 years they have taken a place among the leading subjects in astrophysics, both with observers and theorists. Perhaps I have had a small hand in making them respectable.

Pickering, however, had found how to turn variable stars to use in another way. He saw that much valuable work in this field could be done by the amateur astronomer. The American Association of Variable Star Observers, founded with his blessing and carried on for years by the devoted work of Leon Campbell, channeled its energy into useful fields and was of enormous value, both direct and indirect, to astronomy. It still continues to be so, and it is not the least of his scientific legacies.

When his Directorate ended, Pickering had given the world a broad base of photometry, visual and photographic. He had produced a

The cradle of astrophysics 153

classification of stellar spectra that was to become the foundation of the modern standard, and had presided over the production of the first great catalogue of stellar spectra. Under his direction the modern study of variable stars had begun to expand, to show promise of what it is today. In the course of these explorations he had accumulated that storehouse of scientific data, the Harvard collection of photographic plates. Nearly a million photographs bore a record of the brightness and spectra of stars over the whole sky, systematically accumulated decade by decade. New techniques have increased the accuracy of photometric work, and have vastly extended the range of wavelengths that can be studied, but the photographic plate is not yet superseded. The Harvard plate collection is still an astronomical bonanza – witness its contributions to the study of quasars and X-ray binaries. There are as good fish in that sea as ever came out of it.

So much for ancient history. When I first came to Harvard, the Observatory was at the parting of the ways. I never saw Pickering, never knew Miss Leavitt, though their shadows could still be discerned. I heard tell that Miss Leavitt's lamp was still to be seen burning in the night, that her spirit still haunted the plate stacks. I suspect that some credulous soul (and there were such in those days) had seen me from afar, burning the midnight oil. Shapley had given me the desk at which she used to work.

11

Harlow Shapley

The young director was everywhere, running upstairs two steps at a time, pushing his soft sandy hair off his forehead, greeting everyone with the same casual cheerfulness. He knew exactly what each member of the staff was doing. He made a regular stop at each desk, and with a few well-chosen words (I use the overworked expression advisedly) made each of us feel important. If by chance one of us was not at work when he paid his daily visit, a little note would soon appear there, calling attention to the fact that we were expected to put in regular hours. He had inherited a large and unwieldy institution, one with a reputation to keep up. But he was not to be content with carrying on the Pickering tradition. Adelaide Ames and I used to say jokingly that he had found a Dear Little Observatory, and intended to leave it a Great Institution.

The early success of the Observatory had been made possible by Pickering's astute planning. Women, some of them extremely able, had carried on the bulk of the work. At the turn of the century, work at an Observatory was not only respectable but conferred a certain distinction, and these advantages had compensated for the small pay. This tradition had perforce to be carried on by the new regime. Women still carried out the subordinate work, and very well they did it. But salaries continued to be low; the only alternative would have been to cut back the programs at the moment when the new Director's plans called for expansion. Dr Shapley cynically measured his projects in 'girl-hours'. Part of the compensation was in the form of personal encouragement, which conveyed the sense that everyone's work was of real importance. It was the secret of his early success. More than once I have heard him say: 'I think *I* could do this, so

I'm sure *you* can.' I never knew it to fail. Everyone adored him – the older women and the young girls who were soon added to the team. Adelaide and I called him 'the Dear Director', and soon he was affectionately known as 'the D.D.'. New life was breathed into the institution; what must once have been a routine operation became a happy partnership.

I was in a different position from the other girls. They were employed to do a job, but I was on a Fellowship, so I was independent and had no obligations. When I arrived, Dr Shapley asked me what I wanted to work on, and mentioned that he wished I would prepare to carry on Miss Leavitt's work on standard photometry. But I had other ideas. In my last year at Cambridge I had come to know E. A. Milne, who (with Ralph Fowler) had just published the historic paper on stellar atmospheres. They in turn had been inspired by the brilliant idea with which Meghnad Saha had applied the principles of physical chemistry to the ionization of stellar material, the idea that gave birth to modern astrophysics. Before I left Cambridge, Milne told me that if he had my opportunity, he would go after the observations that would test and verify the Saha theory. When I told Dr Shapley that this was what I should like to do, he promptly opened up to me the riches of the Harvard plate collection. I was alarmed at the thought of handling such priceless material. What if I should break one of the plates? 'Well,' he said, 'you can keep the pieces.' When a little later I inadvertently sat down on a pile of plates, I rushed out of the Observatory and did not dare to return for several days. Fortunately no damage had been done.

The 'D.D.' kept in close touch with what I was doing. It was the first time I had been able to discuss scientific problems freely, and I must have taken up a great deal of his valuable time. He was the most wonderful talker I have ever known. A discussion with him was like a rousing game of ping-pong, ideas flashing back and forth, careening off at unexpected angles and often coming to earth in a breathless finish. He must have found my artless enthusiasm, my unprejudiced ignorance, amusing and stimulating. I had never before been in contact with the raw materials of astronomy, and my questions must have had the revealing quality that one finds in the enquiries of a child, the best of all scientific audiences. I look back on him in those days, 'this way and that dividing the quick mind', with delight and gratitude.

In spite of this vigorous scientific companionship, Dr Shapley kept his distance. He never forgot, or let me forget, that he was the Director of the Observatory. I knew him for more than 50 years, and never once did he call me by my first name. We discussed everything under the sun, I suppose, but he never let our relationship become personal. When I reflected on it – and I often did – I thought that a good description of it was contained in Bernard Shaw's play about the relation between Caesar and Cleopatra. 'Can one love a god?' says Cleopatra when she is asked whether she is in love with Caesar. In fact I once mentioned the parallel to him, but his response was chilly. I only realized later that he had confused Shaw with Shakespeare; he did not know the difference between *Caesar and Cleopatra* and *Antony and Cleopatra*. Alas, I was too much embarrassed to set him right. He need not have worried. It was 15 years before I outgrew my childish dream of playing the Beggar Maid to Eddington's King Cophetua. But of course I kept this to myself. In those days I worshipped Dr Shapley; I would gladly have died for him, I think.

But this did not prevent me from taking a critical view of my idol. He liked to be flattered, and I certainly flattered him. That was one of the reasons why he enjoyed my company. I flattered *myself* that the reason lay in my intelligent conversation. When I noticed that he took equal pleasure in the conversation of some people whom I presumed to think less intelligent than myself, I was jealous, and I am afraid I showed it. I ought to have seen that he mixed with everyone with equal zest. It was part of his natural behavior. 'A *largesse* universal like the sun His liberal eye doth give to everyone' describes it exactly. It was the secret of his personal success.

With his susceptibility to flattery went a less endearing trait. He never forgot or forgave a slight. He was vain and vindictive. A generous supporter, a stimulating companion, he could also be an implacable enemy. In his published recollections he says that he does not remember disliking anyone. I think that was an exaggeration.

Under Dr Shapley the Observatory was expanding in several directions. As a first step it had to come to terms with the University. Up to the end of Pickering's life it had been solely devoted to research, and had had no connection with the teaching of astronomy at

Harvard. Bailey, King and Gerrish had professional titles, but they had done no teaching. Shapley's plans called for a school of astronomy, not only for undergraduates but for graduate students. Here I was the thin edge of the wedge. He soon told me that I should work for the degree of PhD at Radcliffe College, Harvard's sister institution. I was not much interested; I thought that no degree could be higher than the one I had received from Cambridge University (even in those days, before the admission of women, it was only the 'Title of a Degree'). Finally I was persuaded: I need not take any courses (there were none to take), and had only to write a thesis, which I was going to do in any case. One serious obstacle existed: there was no advanced degree in astronomy, and I should have to be accepted as a candidate by the Department of Physics. The redoubtable Chairman of that department was Theodore Lyman, and Shapley reported to me that he refused to accept a woman candidate. I never knew how Shapley handled the problem but by the time I actually received a degree it was in astronomy – the first ever to be awarded at Harvard or Radcliffe. The graduate school of astronomy had come into being.

The Department of Astronomy was of gradual growth. At first there were composite lecture courses, several members of the staff covering their special subjects. Then Harry Plaskett was called from Victoria to build up the organization, and graduate students began to flow in, first from the United States and Canada, soon from every quarter of the world. Gradually the teaching began to be systematically organized. But – it was a puzzling trait – Dr Shapley, apart from an occasional lecture, did not share in it. I think he undervalued teaching as such. More than once I heard him refer to a respected astronomer as 'just a teacher'. Perhaps his attitude infected me. Once I had dreamed of being a teacher, and regarded the teaching profession as the world's highest. But I developed an abiding dislike for lecturing to students. I think that my distaste stems from the inherent conflict between teaching and research. A lecturer must pose as knowing everything about his subject (and some even seem to believe that they do); in research one must continually remind oneself that one knows little or nothing. A scientist who must do both develops a sort of intellectual schizophrenia. Perhaps it was some such conflict that kept Shapley from the lecture room.

Not that he would have been a poor teacher. As a popular lecturer and a public speaker he had few rivals. He could hold an audience spellbound, using all the arts of the actor to enliven the most arid subjects. His timing was flawless. To my own taste he was too much given to levity; he would play to the gallery for the sake of an easy laugh. He believed that he was justified by the results.

It seemed to me that irreverence when speaking of scientific truth was as much out of place as levity in the pulpit. He had shocked me when he gave the memorable lecture in London. A photograph of Messier 8 was on the screen. He pointed out the patches of obscuration. 'Some people might say that they are the fingerprints of God; but perhaps they are only the fingerprints of the careless devil who made the plate.' That was his style, and most of the audience loved it. Some may react to the splendor of the Universe with a wisecrack; to others a dignified silence seems the only possible comment.

12

Stellar atmospheres

Astrophysics, the application of physics to the Universe on the widest possible scale, seemed to provide a laboratory whose scope far transcended the world of the Cavendish. Temperature, density, velocity could be studied over ranges unknown, untouchable on Earth. I saw in the stars a chance to observe phenomena beyond terrestrial scope. Nothing seemed impossible in those early days; we were going to understand everything tomorrow.

Experience has modified that early optimism. Progress has been slow, the road has taken unexpected turns. Often we did not even know what we were studying. We thought we were observing orbital motion; the stars turned out to be pulsating. We thought we were studying peculiar stellar atmospheres; we were collecting information about interstellar material. The history of the gradual understanding of stellar atmospheres reminds me of an idea that had dawned upon me many years before. In the outburst of 'poetry' that had seized me in my schooldays, I had written on the subject of cumulative emotional experience

> ... If all those joys
> Could strike the mind together, sure they would
> Amaze and stupefy, strike blind and slay.

In other words, it is well for us that knowledge comes to us gradually. If we were faced with the full complexity of the facts at the beginning of our search, we should be so bewildered that we might well give up in despair.

Consider the history of the study of stellar atmospheres. Lockyer, the pioneer, recognized a series of stellar spectra, some dominated by the lines of hydrogen, some by those of helium, some by metals or the spectra of chemical compounds. He saw a continuous series, and thought it spelled out the evolution of the chemical elements. Saha, his genius kindled by a knowledge of physical chemistry, understood the series in terms of the Law of Mass Action. Fowler and Milne reformulated the idea in terms of a new concept of astrophysical conditions, at densities far lower than any that had been previously supposed. I was one of those who furnished an observational background for their theory, as I shall presently tell.

But then came disquieting reflections. Stars are not all alike. Hertzsprung and Russell awoke the world to the existence of giant and dwarf stars. Their spectra must show luminosity effects (which Miss Maury had described long before, not knowing what she saw). Some stars were seen to display incompatible atmospheric phenomena: their spectra could not be understood in terms of a single set of conditions. The conception of atmospheric structure emerged (the Sun must have pointed the way to this picture long before), and the theory of model atmospheres was developed. There were other anomalies, too; observations that seemed incompatible with the beautiful uniformity of composition that had been one of the first fruits of the Saha theory. Real differences of composition forced themselves upon us. More recently, ultraviolet, infrared and radio astronomy have brought the stellar atmosphere and envelope to a stage of extraordinary complexity. After more than 50 years, the outer layers of the stars are full of mysteries and surprises. The growth of the concept of stellar atmospheres has been slow, mercifully slow. I saw the curtain rise upon that drama, but it is still being played out, and I do not think we have witnessed the final *dénouement*.

The Observatory had two great libraries: one contained the printed word, the other the photographic records of the stars and their spectra. I applied myself first to the former, and it is a course that I recommend to every student. Through the library I went, shelf by shelf, arranging its contents in the pigeonholes of my mind. There were the journals (how much fewer than today!), each with its own specialized content. Equally important were the Observatory publications, and here one learned where the dry bones of astronomical knowledge were stored, bones to be assembled and clothed with the

Stellar atmospheres

flesh that would present the stars as complete individuals. That was a new idea at the beginning of the century; position, motion, color and spectrum were separate facts, not yet welded into the study of stellar personalities.

Next I turned to the plate library. The *Henry Draper Catalogue* was not yet completely off the press, the first pigeonholing of 250 000 stellar spectra. Saha and Fowler and Milne had brought the spectral sequence into a physical order for the first time, but only in the most general terms. I followed Milne's advice, and set out to make quantitative the qualitative information that was inherent in the Henry Draper system.

A few weeks after I had begun to examine the spectra of the stars for the first time, Shapley called me into his office. I found him looking rueful and apologetic. 'Menzel has come,' he said. Henry Norris Russell had sent his student, Donald Menzel, to obtain material for his doctoral thesis (on the very same subject as the one I had chosen) in the Harvard plate collection. Russell, who was to have a great influence on my own work, had of course seen the enormous possibilities of the Saha theory, and it was natural that he should direct his ablest student to the fountain-head of the facts on stellar spectra.

Here a great mistake was made, a great opportunity missed. This should have been the occasion for a spirited and fruitful collaboration, for Donald Menzel had been trained at Princeton with a basic knowledge of laboratory spectroscopy. I, on the other hand, knew almost nothing about spectra, for the training at the Cavendish had stressed the physics of the nucleus, rather than the external properties of the atom. (I had heard Rutherford refer to the latter in disparaging terms as 'descriptive botany', and the phrase grated on me, for I still had a respect for descriptive botany: the study of systematics has left a mark on my approach to the stars.)

Shapley could have brought Menzel and me together as friends and fellow-scientists. But that was not his way. He operated in the spirit by which Kipling described the British rule in India: 'Strict supervision, and play them off one against the other.' Not for over 20 years did I realize that he was practising this technique. Not until Donald Menzel succeeded him as Director of the Observatory did I see that we had been deliberately kept apart. I was never to be a Director –

my sex debarred me (and I am grateful for that). But if such had been my fate I would have made it my business to promote collaboration and not to fan hostility. How could I have been blind for so long to these divisive tactics? I looked on Shapley in those early days with uncritical adoration; perhaps, with my natural tendency to masochism, I subscribed to his idea that one does one's best work when one is miserable. 'Work', he used to say, 'kills the pain.' I ought to have remembered the precept inculcated at the religious school: 'The words "thy will be done" should not suggest a mourning widow, but a young person rushing out into the world to do good'. It would have served me better than Shapley's gloomy philosophy. He used to say that no one could earn a Doctor's degree unless he had suffered a nervous breakdown in the process. I told him stoutly that this was all wrong. He rejected one excellent candidate for the PhD on the grounds that the latter was not a serious person – he was always laughing. 'Well', the boy replied, 'It's better than weeping.' He told me this many years later; he was a loss to astronomy.

It was my first taste of professional jealousy, the struggle for priority. I had much to learn. The young are in too great a hurry, too eager to be the first to get the credit for a new discovery, a new idea. I have come to feel a nostalgia for those early days when an astronomer was content to accumulate the basic facts, and knew (as the observer of double stars still knows) that it is not for him to reach the final interpretation. Only gradually have I come to appreciate that '*l'homme, c'est rien; l'oeuvre, c'est tout*'. I used to think: 'This is *my* problem'. I guarded it jealously; I snarled at anyone who dared to approach it. I have come to know that a problem does not belong to me, or to my team, or to my Observatory, or to my country; it belongs to the world. I should like to be remembered not for any observation, any idea, but for what I immodestly call 'Payne-Gaposchkin's Principle', the criterion by which one's work should be examined. Am I thinking of myself, or of the advancement of knowledge? Professional jealousies, struggles for priority, wither before that question.

But I pressed on alone. It was clear that some quantitative method must be devised for expressing the intensities of spectral lines, and I set up a crude system of eye estimates. Next came the identification of the line spectra, the selection of known lines for examination, and the arduous task of estimating their intensities on hundreds of spectra. The analysis of laboratory spectra was in its infancy, and I think that the pioneer work of Henry Norris Russell in this field was his greatest contribution to astrophysics.

There followed months, almost a year as I remember, of utter bewilderment. Often I was in a state of exhaustion and despair, working all day and late into the night. Every evening I would listen for Dr Shapley's light step, as he ran across from his residence to the Brick Building. He would stop at my desk and speak the words of encouragement that were the breath of life to me. Once, after I had toiled for many months, he asked me whether I did not think I ought to publish something, to give some evidence of the work I was doing. 'No', I said, 'I haven't solved the problem yet. I should regard it as a confession of failure.' He accepted it, and I think he was pleased.

Finally some light dawned in the darkness. The intensities of the lines of silicon, in four successive stages of ionization, made some sense at last, and I made my first determination of the temperatures of the hotter stars. He encouraged me to write my first paper on stellar spectra. I wanted to sign it C. H. Payne, but he insisted on the full name: 'Are you ashamed of being a woman?' he asked. Silicon had broken the ice for me; it is still one of my favorite atoms. Now I saw my way clear.

This narrative would not be a true one if I mentioned only the successes. There were many failures, many blind alleys. I think I am an expert on 'How Not to Do Research'; indeed it might serve as a title for this book. Let me record one such failure. Soon after I came to Harvard, Shapley acquired a recording microphotometer, and the study of the distribution of energy in stellar spectra was begun. Dr Shapley set the work in motion, and soon the first records of the spectra of A stars were hanging in the darkroom to dry. I could not resist them. I noted with interest that the continuum dipped between the lines Hγ and Hδ, and remarked to Adelaide Ames that this was evidence of a departure from a Planck distribution. Perhaps it was an absorption band. I think the spectrum was that of Vega. She passed on this remark to the Director and his reaction was slightly rueful: 'What else did Miss Payne see?' At that I felt I should discuss it with him myself, especially as I had taken a look at his material even before he had had the chance to see it for himself. I went to him, full of ideas. He was even fuller of ideas than I. Neither of us reflected that the spectrum was uncalibrated; calibration, that most essential and difficult of all processes, was to come later. The enchanting idea arose, between us, that the dip in the spectrum was caused by high-velocity iron atoms falling on to the surface of the star, perhaps from comets or meteors. We examined other spectra of A and B stars; the dip was always there. With the enthusiasm of youth we published a joint paper on the infall of meteors into stars, almost the only collaboration we ever made.

Then came disillusionment. Professor King, punctilious photometrist that he was, pointed out that the dip was not real at all. It was produced by the absorption in the heavy glass prism and in the lens itself. He went on to determine the extent of the dip for all the instruments and prisms in use at Harvard. I was terribly chagrined, and so, I am sure, was Dr Shapley. He promptly published a correction and retraction, but refused to allow me to sign it too. 'You only want to do it out of loyalty', he said. This was my first lesson in How Not to Do Research, my first introduction to the necessity for calibration.

Disillusionment was rarer than exaltation. As the work progressed, and the relation between line intensity and temperature began to take shape, the ionization potentials of the atoms took on great importance. In those far-off days, the analysis of laboratory spectra was only at the beginning, and this quantity was unknown for many of the metallic elements, even for iron. It occurred to me that one could infer the ionization potential of an atom from the behavior of its lines in the spectral sequence, when the scale had been established by means of other atoms of known properties. Donald Menzel hit on the same thing at about the same time – the idea was in the air – and published it too. Would that we had put our heads together! But this was not to be. When I brought my result to Dr Shapley his enthusiasm was kindled: I must write it up at once and send it to the *Proceedings of the National Academy*. But the deadline was next day. 'Write it up at once!' he exclaimed, 'and I'll type it for you.' What a glorious evening! I wrote, he typed, far into the night. Into the mail

Stellar atmospheres

it went. And I walked back to my room in the dormitory in a dream. My feet did not seem to touch the ground. 'I never knew before', I thought, 'what it means to walk on air.' Never again have I experienced that feeling; it was almost like flying. I had not wanted to tell him that I was quite a good typist myself. When still at Cambridge I had accustomed myself to compose on the typewriter. But he did not find that out until much later.

Two years of estimation, plotting, calculation and the work I had planned was done. I had determined a stellar temperature scale and had measured the astrophysical abundance of the chemical elements. It was to be a PhD thesis (though, as I have said, I cared little about the degree). Dr Shapley said I must make it into a book, which I wrote, in a kind of ecstasy, in six weeks. It was called *Stellar atmospheres*, but I added a sub-title: *A contribution to the observational study of matter at high temperatures*. Two years earlier I had learned that this was to be the subject of the Adams Prize Essay at Cambridge, and that it was aimed at Ralph Fowler (who indeed received the Prize). I wanted to show that I had my contribution to make to the topic of the hour. Probably nobody even noticed that hubristic sub-title.

Two comments on this work come down to me on the winds of memory. When I met the great Alfred Fowler, whom I regarded as the greatest living spectroscopist, he looked at me quizzically. 'Miss Payne?' he said drily. 'You're very brave.' And Eddington, in reference to my little book, commented that my ideas were 'not so wild as might at first appear'.

When I returned to visit Cambridge after I had finished this first essay in astrophysics, I went to see Eddington. In a burst of youthful enthusiasm, I told him that I believed that there was far more hydrogen in the stars than any other atom. 'You don't mean *in* the stars, you mean *on* the stars', was his comment. In this case, indeed, I was in the right, and in later years he was to recognize it too.

Two years of daily contact with stellar spectra had brought one idea to the front. The composition of the stars was amazingly uniform. The years have modified this conclusion, but even today I find the uniformity more striking than the diversity.

My days as a student and a free agent were over. Fellowships had

Fig. 8. Members of the staff at Harvard College Observatory, 1925. From left to right: Harvia H. Wilson,
Agnes M. Hoovens, Antonia C. Maury, Ida E. Woods,
Annie J. Cannon, Mary Howe, Margaret Harwood,
Evelyn F. Leland, Arville D. Walker, Lillian L. Hodgdon,
Cecilia H. Payne, Edith F. Gill, Margaret L. Walton,
Mabel A. Gill, Florence Cushman

supported me for two years. Now I must find a place in the world. Dr Aitken, Director of the Lick Observatory, asked me to go there and take a Research Fellowship. What would my future have been, I wonder, if I had accepted? When I told Dr Shapley, he was indignant. Dr Aitken, he said, should have consulted him first. He offered me a position at Harvard Observatory. In my innocence I did not ask how much he was going to pay me, or realize how little it would be. Nor, I think, should I have cared very much. I accepted the offer, and settled down to work at Harvard College Observatory, which was to be my home for more than 50 years.

13

Spectra and luminosities

After two years in the United States I knew that I wanted to stay there. The Harvard Observatory, with its heady atmosphere of scientific discussion and progress, satisfied my most urgent desires. I was too young as yet to suffer from nostalgia.

I quickly found out the difference between a Fellowship and a job. The former pays at the beginning of the month, the latter at the end. Suddenly I found myself without funds. Imagine my embarrassment. I was too proud to tell anyone how naïve I had been. So I tided myself over by pawning my violin and my jewelry. Afterwards it seemed laughable, but I had known humiliation and panic, and ever since I have felt sympathy for those in a similar situation. When I had students of my own I learned to recognize the signs.

For two years I had lived in a dormitory, but now I was able to have a home of my own. The feminine urges to cook, to sew, to entertain were natural to me, and I have continued to enjoy fulfilling them. When one is spending several years in bringing a project to fruition, there is great satisfaction in producing a masterpiece in the kitchen in a couple of hours. I had once pictured myself as a rebel against the feminine role, but in this I was wrong. My rebellion was against being thought, and treated, as inferior. But being inferior is not the same as being different. Of course women are different from men. Their whole outlook and approach are testimony to it. 'All nature', said Swedenborg, 'is masculine; the Earth is the Mother.' That symbolical summing-up puts it better than I can do. All that I have done is respond to the quickening influence of the Universe. And during my whole scientific life I have never had a sense of being inferior, never been conscious that my fellow-scientists considered me inferior, on account of my sex.

The position that I held at the Observatory was a very indefinite one, and I never succeeded in getting my duties defined. Dr Shapley suggested that I follow up *Stellar atmospheres* with another book about supergiants, and he also asked me to undertake the editing of the Observatory publications. No restrictions were placed on anything else I felt like doing.

The supergiants came straight out of the *Henry Draper Catalogue*, simply by dint of making a list of the stars mentioned in the Remarks as having sharp lines. They were, of course, a gold-mine, as Shapley had known they would be. The way had already been pointed by Hertzsprung's demonstration that Miss Maury's 'c-character' was associated with high luminosity. I remember writing *The stars of high luminosity* without pleasure. My first book had arisen naturally from a quantitative survey of the spectral sequence; this one was a predetermined study that had to be made to fit the title. Moreover, the superficial simplicity of my first treatment had become blurred. Density, as well as temperature, differed from one stellar atmosphere to another; the physical picture was becoming confused, and the confusion has increased from that day to this. Time and again I rebelled – it was not a coherent book, and I would not write it! But authority prevailed, in more ways than one.

The supergiants led inevitably to studies of galactic distribution and structure. I was brought back to the conviction, reached at Cambridge several years earlier, that allowance must be made for interstellar absorption. The known colors of the stars, exiguous as they then were, pointed inexorably in that direction. But Dr Shapley would have none of it. His whole picture of the Galaxy, the structure of the system of globular clusters, seemed to hang on the issue. I record with shame that I allowed myself to be persuaded, to adduce evidence of faint blue stars at low latitudes, and to express the opinion that interstellar absorption was at most a minor factor in the study of stellar distribution. This is one of the exhibits in my gallery of How Not To Do Research. It set my own thinking back several years, but fortunately other astronomers were wiser, and Trumpler's study of the distribution of open clusters established the importance of interstellar absorption, especially in low galactic latitudes.

Another addition to the gallery arose out of my study of the hotter stars. While still a student at Cambridge, I had asked Milne why the Stark Effect was not recognized in stellar spectra, although the Zeeman effect was well known in sunspots. I was sure that the Stark Effect would be detectable if it were looked for in the proper place, and the proper place seemed to be the B stars. Armed with T. R. Merton's laboratory studies of the spectrum of helium, I traced the profiles of the lines in the spectra of the hottest stars and detected the characteristic asymmetries for the line at 4471 in a number of spectra. I believed that the data were convincing (as indeed I still do), but when I brought my work to Shapley, he was sceptical, and his scepticism was backed up by the weighty opinion of Henry Norris Russell. I was not to publish my results.

Years later, long after Otto Struve (in collaboration with C. T. Elvey) had established the existence of the Stark Effect in stellar spectra, I told Struve that I had noted the effect in 1925. With characteristic generosity he said that he would mention my prior discovery in the history of astrophysics that he was writing. But I would not have it. I was to blame for not having pressed my point. I had given in to Authority when I believed I was right. That is another example of How Not To Do Research. I note it here as a warning to the young. If you are sure of your facts, you should defend your position.

The study of the high-luminosity stars opened up two vistas that have been expanding before my eyes ever since. The first was the grand array of Wolf–Rayet stars. Never shall I forget my first sight of the spectrum of Gamma Velorum! Among stellar spectra the Wolf–Rayet stars remain my first loves. Forty years later, when I was asked to speak at the opening of a conference on these still enigmatic objects, I likened them to the figure of the Nymph on the Grecian Urn: 'Forever shalt thou love, and she be fair'. Still fair, still elusive, the Wolf–Rayet stars continue to attract and puzzle us. I first met that great astronomer J. S. Plaskett at the 1925 meeting of the International Astronomical Union; he had just published his classic paper on their spectra. How did I summon up the courage to defy him in public, and maintain stoutly, in the face of his opposition, that some Wolf–Rayet stars show absorption lines? It was like a puppy defying an elephant.

With what intense delight did one use Edlén's analysis of the spectra of carbon, nitrogen and oxygen to identify most of the bright lines in the Wolf–Rayet spectrum for the first time! With even greater joy I have witnessed the gradual growth of knowledge: the recognition that many of them are spectroscopic binaries, and the culmination when (many years later) my husband was the first to show that many of them are eclipsing stars and to deduce their physical properties for the first time! Many of the problems that were opened up 50 years ago have been solved today, but the Wolf–Rayet stars still have some surprises for us.

A second vista, even wider, opened up in the shape of the variable stars. In those distant days all variable stars seemed to be of high luminosity. But they were regarded rather in the light of second-class citizens of the heavens. Shapley did not hesitate to call them pathological stars. They are recognized today as the representatives of definite stages of stellar development, but in the 1920s there was a tendency to think of them as the province of the amateur.

This, of course, was quite unjust. Miss Leavitt's grand discovery of the period–luminosity relation, followed by Hertzsprung's flash of genius in using the Cepheids as standards of absolute luminosity, had raised them to a high place, and paved the way to Eddington's theoretical study of stellar pulsation. Bailey's work on the RR Lyrae stars in globular clusters had raised them, too, to 'professional status', and had paved the way for Shapley's epoch-making transformation of our picture of the Galaxy.

A survey of variable stars as an approach to the structure of the Milky Way was one of Shapley's grand projects. It was conceived on a scale so vast as to prove impractical. Several hundred fields were to be photographed repeatedly, variable stars were to be discovered, and used to trace the structure of the Galaxy. Interest was concentrated at first in low galactic latitudes, especially in the direction of the galactic center. The work was placed in the capable hands of Henrietta Swope, who came to the Observatory two years after I did, and evinced an extraordinary flair for discovering variable stars.

Two results emerged in a short time. The search for variables produced so many, even in a few areas, that the task of complete coverage was seen to assume colossal proportions. It also became clear, even to Shapley (who, as I have mentioned, had had other ideas) that interstellar absorption plays a major role, at least in low galactic latitudes. Variable stars could indeed be discovered, their brightness determined and their types and periods found, but their distances could not be directly deduced, and the galactic structure remained elusive. Emphasis was therefore shifted to high-latitude fields, especially those in the southern hemisphere near the longitude of the galactic center. Many areas in the northern Milky Way and close to the galactic plane are still not completely explored, and in these areas our picture of galactic structure remains confused and incomplete.

For the study of variable stars, even if their types of variation are established, is not enough to define galactic structure. The fundamental problem is a photometric one, as Pickering had realized several decades earlier when he assigned to Miss Leavitt the task of determining standard magnitudes. The impact of photometry is twofold: colors as well as magnitudes are essential if the interplay of distance and absorption is to be disentangled. Only the resources of modern photoelectric photometry, when applied to variable stars in low latitudes, could have produced the result that Shapley had in mind when he planned his survey of Milky Way variables.

In the 1920s the requirements seemed to be more modest. Except for the North Polar Sequence and the brighter stars in a number of standard regions, there were few reliable photographic standards. Over most of the sky there were no such standards fainter than the thirteenth magnitude. Without them the survey of Milky Way variables could be no more than qualitative. With this in mind, Shapley recurred to his original suggestion that I should turn my attention to the determination of photographic magnitudes and standards. Alas for my beloved spectra! It was hard to leave them, and to turn to the arid field of standard photometry. But such was my devotion to the Director that I did not refuse him, and I embarked on an endless undertaking. Photographic, photovisual and photo-red standards were to be established. As it turned out, the undertaking was to be sterile too, for the new photoelectric techniques had begun to usurp the field before it had produced any fruits.

Standard photometry is one of the solid foundations of astronomy. Its requirements are as rigorous as those of positional astronomy and of the measurements of radial velocity. Like these, it has been the lifetime study of some of the ablest astronomers. The North Polar Sequence is a monument to F. H. Seares and to Miss Leavitt. I well remember the day when an international committee proposed to supersede the North Polar Sequence by a set of photoelectrically determined standards, and the quiet indignation of Seares at the thought that the tremendous work involved in the photographic standards might be set aside.

I do not think that I could have made good contributions in the field of standard photometry. It calls for qualities that I do not possess. If my work has been of any value, that value has consisted in bringing together facts that were previously unrelated, and seeing a pattern in them. Such work calls for memory and imagination, neither of which has much to contribute to standard photometry. I am a field naturalist, not a surveyor. I wasted much time on this account. This was another lesson: loyalty and devotion are not sufficient to qualify one to do research.

My change in field made the end of the decade a sad one. But it led to another activity that would occupy me for many years. I began to explore the potentialities of the Harvard photographs, not for the discovery of variable stars, but for the study of variables already known. The impetus was provided by a published catalogue of stars suspected to be long-period variables (Mira stars), about which little was known save their variability. The search led to some surprising results: one of the subjects, for example, turned out to be an eclipsing star. We determined the elements of no less than 100 previously unstudied, bright long-period variables. Thus I came upon a lode that was in future years to yield much precious metal, and occupy me, off and on, for the rest of my scientific life.

14

Editorial experiences

To be the Editor of the Observatory Publications was a great adventure. I loved the material side of it – the hurly-burly of the University Printing Office, the look and smell of the fresh galleys, the detail of proofreading, the craftsmanship of makeup. The Printing Office handled only the typesetting. I edited the papers, drew the diagrams, read the proofs, assembled the pages. There was a lot to learn, and all of it was invaluable experience. Errors were unforgivable, to be atoned for with much mental anguish.

The intellectual side of editing was harder to master. Shapley (scarred, perhaps, by the strictures of his one-time professor) warned me not to make the Harvard publications read as though they were all written by the same person. That, he said, had been true of the Mount Wilson publications under the editorship of F. H. Seares.

It would have been difficult to impose uniformity on the authors of the Observatory publications, who seemed to me to be a group of stubborn individualists. My standards of style were rigorous. I remember one contribution, which I carefully weeded of colloquialisms and provincialisms and returned to the author. It came back to me with all the original wording meticulously restored. I was young and the author was old; I capitulated. I fear that I spared Dr Shapley as little as I did my colleagues, as when he used 'uxorious' when he meant 'usurious'. But I defended him too. When one of the said colleagues (smarting, I suppose, under some of my preciosity) pointed out that the Director had used the word 'finitude', which was not in the dictionary, I retorted that if it wasn't, it ought to be.

I am afraid I presumed upon my position. There came a day when I wrote, sent to the printer and actually published an unsanctioned note that had not received the imprimatur. How could I have committed such an effrontery? Swift was the descent of wrath, and richly did I deserve it. I became jealous of my position as Editor, and felt that I had a veto on content as well as style. Shapley must have noted this, and who can blame him if he resented it?

A case in point was the appearance of *Harvard Circular* no. 350: 'The supergalaxy hypothesis', which Dr Shapley published in 1930. The paper had gone quickly through the press: I had not seen the manuscript or read the proofs. Often I regretted that my hearing was so acute that I could hear much of what was said in the corridor, especially when (as in those days) all the office doors were supposed to be kept open. 'Miss Payne', I heard him say to someone, 'will be furious if I don't show it to her.' Even if not furious, I was at least curious. When I read the paper I was appalled: I was sure that it was all wrong – that I could prove from my own knowledge that he was mistaken. He must have sensed that, and deliberately kept it from a critic. To this day I cannot imagine what inspired him to write it, if not some impish urge to fly in the face of the contemporary trend of thought. Lindblad had just published his classic paper on the rotation of the Galaxy. Shapley's paper was never followed up, as far as I know, by its author or anyone else, and I have never seen it referred to.

For 20 years I edited the Observatory publications. This included not only the *Bulletin*, *Circular* and *Annals*, but the books which appeared under the title of Harvard Monographs. One of my larger tasks was the assembling and editing of Shapley's own book on *Star clusters*. Here I acted as assistant, too: I assembled the references and drew the diagrams. This was my first introduction to the subject, the kindling of an interest that has grown with the years.

As time passed and the output grew, editorial work occupied much of my time. Foreign visitors added to the burden, for their work called not only for editing, but often for complete rewriting in English. I had already experienced the problem myself in a small way, slowly mastering the subtle differences between English and American spelling and idiom. I fear I have not yet mastered them completely. Even so small a matter as the placement of the decimal point is distinctive. I recall that I mentioned to Frank Schlesinger (a great man with whom I never established a happy relationship) that in England the decimal point is placed at the top of the number, in America at the bottom. He denied it, and only agreed after reference to the current number of the *Monthly Notices*.

Editing taught me much. An author has an instinctive love of his own words, even when they are repetitive, ambiguous, or simply incorrect. In the beginning I attempted to eliminate not only these errors, but the less indefensible solecisms and the departures from pleasing rhythm. Time has curbed these critical impulses. People should be allowed to express themselves in their own words. If their thinking is muddled, their expression will be muddled too, and an editor who impresses coherence on an incoherent argument ends by conferring unjustifiable authority on the presentation. Good scientific thought has an esthetic perfection, which reveals itself in the language of presentation. When a person writes of what he understands in every detail (and only then should he write about it) his words acquire a natural flow, and he says exactly what he means.

Tabular matter is often an editorial nightmare. I look back on a multitude of unheaded columns, ill-defined and incorrectly defined terms. And I should like here to issue a plea that data published for the first time should be clearly distinguished from those derived from other sources and included in the same table. If it had no other effect, this practice would prevent the proliferation of errors, which often descend unwittingly from one publication to another.

An editor comes face to face with the fierce insistence on authorship and priority. These conflicts often lead to premature and incomplete publication, so that today one may find an important result spread over several brief and partial papers. I have come to wish that all scientific work could be published anonymously, to stand or fall by its intrinsic worth. But this is an unrealistic wish, and I know it.

Allied to this problem are those raised by the refereeing of scientific papers. I cannot help wishing that every contribution that bears the mark of sincere and responsible work might see the light of the printed page. But this is also an unrealistic wish, as every editor knows. We cannot escape some system of critical scrutiny. Referees are traditionally anonymous; I wish that the authors of refereed papers might be anonymous too. Many papers, of course, bear the unmistakable marks of their origin. One can say without hesitation: '*Ex ungue leonem*', or, on occasion, '*vulpem*'. But some do not, and anonymity would ensure that the work of an unknown author would receive serious consideration – it *might* be the work of a 'Great Man'. The young are notoriously controversial, and often they are right. This practice would give them a fair field and no favor.

After 20 years I felt that editing was occupying too much of my time, and succeeded in shaking the burden off. But I am grateful to him who first placed it on my shoulders. As Shapley said, I had had my finger on the pulse of research. I learned much, for which I cannot be too grateful. But I learned, too, that nobody loves an editor.

15

Visiting astronomers

There is no need to travel in order to sample the currents of astronomical thought. Harvard Observatory was the crossroads of the world in Shapley's early days. Visitors flowed through – some to make brief visits, more to lecture and work on the Harvard photographs. Let me conjure up a few personal memories: to speak of the work they did would carry me too far afield.

The first flurry of excitement came soon after my arrival. 'Russell has come!' was the word that went around. We young people put aside all work and sat at his feet. Henry Norris Russell was a formidable figure, tall and lean, endlessly voluble, speaking with the voice of authority. One sat for hours listening to a torrent of information. At his first visit he was full of Catalán's new recognition of multiplets in laboratory spectra and all the new vistas that it implied. Nobody else ventured to say anything; we drank until he ran dry. After several hours would come a time when his words flowed more and more slowly. Finally one would hear him murmur: 'Mustn't go to sleep', and then lapse into brief catnaps, punctuated with more words of wisdom. The joking remark that he was the only scientist who had been known to go to sleep during one of his own lectures was literally true; I have seen him do it.

Russell had been Shapley's teacher and mentor, and his word was law. If a piece of work received his imprimatur, it could be published; if not, it must be set aside and its author had a hard row to hoe. His word could make or break a young scientist. It was my good fortune to receive the stamp of his approval in the beginning, though he vetoed some of my cherished ideas. I respected and feared him, but I could feel no affection for him. Russell's visits were social occasions too. The Shapleys would invite the whole staff of the Observatory to a party, at which the guest of honor was again the chief, if not the sole performer. He had a capacious memory, and would recite serious verse (Milton, William Cullen Bryant) for what seemed like hours. Then with a change of pace he would declaim a long string of limericks (all, of course, most discrete, as befitted his audience). Or he would organize and play the chief roles in a series of charades; I remember him sitting on the floor, swathed in a sheet, enacting the Death of Archimedes. Or playing the Pope at the Coronation of Napoleon, the latter part being taken by Shapley, who fancied (quite correctly) that he bore a resemblance to that hero. The whole celebration was a gigantic family party, with Russell the *Paterfamilias*.

Shapley and Russell were very different characters. Russell was not sensitive, vain or vindictive. His was the certitude of conscious superiority. It did not, I think, ever occur to him that he might be exploiting those who worked under him. He took R. S. Dugan, Charlotte Moore Sitterly and Newton Pierce for granted. And yet how much of the work for which he is remembered is due to them! He was a great man, but he had not Shapley's gift of letting his associates share the sense of achievement.

Another visitor came to the Observatory in those early days, and he came not to teach but to learn. Ejnar Hertzsprung was expected. Dr Shapley was in a panic: someone must check all the labels on the cabinets in the plate stacks; if there was an error on them, Hertzsprung would certainly find it! All day he would sit estimating magnitudes, recording his estimates in tiny, exquisite handwriting. He left his mark on our work on variable stars: the 'step-method' of expressing brightness, and the system of studying periodic variables by determining the reciprocals of their periods, rather than the periods themselves. Also we learned from him the pitfalls of spurious periods; he himself had long ago been fooled by VV Orionis, but the lesson is a hard one to learn. He taught me much, for though he hated to waste time, he was always ready to discuss a scientific problem. He was the very model of a dedicated scientist, always eager to further knowledge, but scarcely troubling about claims of priority. The color–magnitude array, the mass–luminosity relation, supergiants, the applications of the period–luminosity relation, the motions of the Magellanic Clouds, these are but a few of the forward steps in which we can trace his footprints.

Once I had the temerity to invite him to luncheon at my apartment. I recall the anxious care with which I prepared the meal. As soon as the last crumb was consumed, he rose briskly to his feet. 'Now', he exclaimed, 'we can go back to work!'

In after years our paths were often crossing. His eyes were always on the stars. The last time I saw him was at the celebration in honor of his ninetieth birthday. In my contribution on that occasion, I spoke of the work my Husband and I had just completed on the variable stars in the Magellanic Clouds. I shall always cherish the memory of the gleam in his eye, and the gusto with which he spoke: 'It is good to hear of someone making *half a million* estimates'; I could see that we had done a work after his own heart. He could have paid no greater compliment.

In the early days of astrophysics it was natural that Meghnad Saha should visit us. He left a clear picture of a great and simple personality. He told me that his whole life had been changed as a result of reading *The cloister and the hearth*. Before that time he had believed that a celibate life was best; but the book had resolved him to marry. How strange that a Victorian author should so influence an oriental! On another occasion he mentioned that he had better not die in the United States. Mystified, I asked why not? He replied, 'Because I am a Parsee. My body must be placed on a tower for the vultures.' His basic faith had not been touched by Charles Reade.

The young astrophysicists came to visit too. Albrecht Unsöld and E. A. Milne had been conducting a spirited controversy on the printed page, no holds barred. They had never met. It chanced that their visits to Harvard coincided. One day, when Milne was talking to me in my office, Unsöld walked in, and it was my fate to introduce them to each other. They stood there, Milne small, delicate and fair; Unsöld tall, dark and rather forbidding. There was a long silence. Then someone began to laugh; the gesture was contagious, and soon we were all laughing till the tears ran down our cheeks. Friendship blossomed, we went on a mountain-climbing expedition. Unsöld, accomplished mountaineer, strode to the top of Mount Chocorua, but Milne and I, daunted by the granite slopes, remained in the lower regions.

A longer visit was paid by Boris Gerasimovič, who came from Pulkova to study the spectra of bright-line B stars. I had much to do with him in my editorial capacity, and putting his papers into English was an education. His considerable contributions to astrophysics have been undervalued; he was a great man, crippled by the accidents of history. I found Boris Petrovich rather alarming; I had no conception of the terrible background that lay behind his sardonic humor. But I spent many an hour in the little apartment where he lived with his wife Olga Michailovna. My friendship with them was to play a crucial part in my future, but I had no inkling of it then.

Knut Lundmark came to lecture, another great man whose work has not been sufficiently appreciated. He had an insight comparable to that of Hertzsprung, and was the first who perceived the significance of supernovae. A memorable lecturer, he lent a touch of portentousness to his pronouncements. This was the more delightful when he interlarded them with slang phrases (picked up from Frank Bowie) uttered in oracular tones.

I hesitate to speak of Otto Struve for fear that anything I can say will not do him justice. He came to Harvard to study the interstellar calcium lines on the spectra in the photographic collection. As always, he worked unremittingly day and night, and scarcely talked to anyone until he had completed his work. But then, I have the memory of spending a whole night in astrophysical discussion, the day before he had to leave. On another occasion we met at a meeting of the Royal Astronomical Society in London, and afterwards we walked right across London, talking, talking, talking. I saw him often after that, at Yerkes Observatory, when he was Director, at MacDonald Observatory, at Green Bank. Always the dedicated scientist, always at work, he has left a memory of inflexible purpose.

I have left until last the visitor who meant the most to me. Svein Rosseland came as a visiting lecturer, and I think his teaching was above the heads of his listeners. To me he was not only a fellow-scientist but a valued friend. I was received into his little family, and there I found human companionship such as I had never known, sympathy and understanding and a place where I could speak my own thoughts. The Rosselands have a unique place in my affections.

Many years later, when I was married and had a family, he came to visit us and we sat talking in the garden. It was 1939 and the clouds of war were gathering on the horizon. After he had left, we noticed that he had left his hat behind and taken my husband's by mistake. It is not part of my story to tell about what he endured – the occupation of Norway, the escape to Sweden, the long journey across Russia, the crossing of the Pacific at the time of Pearl Harbor. Once more in the United States, he came to visit us again. 'Last time I was here', he said, 'I took your hat by mistake; here it is.' We were able to return his own to him, and I still cherish, as a historical relic, the hat that encircled the globe.

16

At the cross roads

My salad days were over. I had spent my first few weeks at Harvard in a systematic survey of the library, noting and arranging in my mind the various sources of information on stellar astronomy. I had spent my first few years there in ranging over the astronomical photographs, collecting and classifying the celestial flora. I choose the word deliberately, for my approach was that of the field naturalist. Just as I had learned the intricacies and pitfalls of systematic botany 15 years earlier, I had spent almost a decade in identifying and pigeon-holing the astronomical species.

The study of botany had led inexorably to the study of plant physiology and the forbidding intricacies of biochemistry. I had given up the subject as too difficult; astronomy, I hoped, would prove to be a simpler science. But systematic astronomy had led to the problems of the energy sources and development of stars. The chemistry of the atomic nucleus was just as basic to astrophysics as biochemistry was to the study of living things, but it was equally inscrutable.

I saw this branch of knowledge develop almost from its beginnings. Eddington had speculated that energy might issue from the atomic nucleus at a temperature of many million degrees, just as steam issues from water at its boiling point. Russell had spoken of 'giant-stuff' and 'dwarf-stuff' as the still unidentified sources of the energy of large and small stars. Then had come the discoveries of Bethe and von Weizsäcker, of Salpeter and Öpik, and the sources of stellar energy and the course of stellar development were clarified.

I had done my part in the spade-work of arranging the data of observation in a reasonable and understandable order. I had turned reluctantly from the interpretation of stellar spectra to the less glamorous but equally important work of standard photometry. Now I wanted to spread my wings and get a glimpse of what was being done elsewhere.

In 1930 I was able to realize the long-indulged dream of visiting the great observatories of the West. When I was still working with stellar spectra, Shapley had talked to me of the possibility of acquiring an ultraviolet spectrograph comparable to the Crossley instrument at the Lick Observatory, and had suggested that I go there and familiarize myself with it. This was in fact an unrealistic and optimistic idea, considering the New England climate and my own very limited knowledge of instrumentation. Fortunately nothing came of it. But I had not forgotten it, and when my plans to visit the West took shape, I reminded Shapley that I would look into the possibilities of the spectrograph. '*What* spectrograph?' was his crushing rejoinder. For by that time Harry Plaskett had made plans for a very different instrument – one, incidentally, that was not to produce results. Undeterred by this setback, I carried out my plan to visit California. Frances Wright had come to work with me at Harvard a couple of years before, and we had found many common interests. Together we set off on an adventurous drive across the United States, by routes and over roads very different from the superhighways of today. Between Kansas and California there were only dirt roads.

That visit to Mount Wilson Observatory is something to remember. I was received with the utmost kindness and hospitality. I met the legendary figures of Western astronomy: kindly old Dr St John, the gay, ebullient Adriaan van Maanen, Frederick Pease, Edison Pettit, and the distant, forbidding Director, Walter Sydney Adams. I received a flattering and exceptional offer of observing time. They had expected that I should want to use the spectrographs. But some masochistic impulse reminded me that I was now dedicated to photometry, so I spent the precious hours on the mountain on color indices for the twin clusters in Perseus.

I felt a special reverence for Dr Frederick Seares, then the prestigious Editor of the *Astrophysical Journal*, for he had been Dr Shapley's teacher. This was an opportunity to learn about photometry from one of the makers of the North Polar Sequence, as well as to discuss the duties of an editor. I fear I took myself very seriously.

Even more impressive were the interviews granted by the great Edwin Hubble. He spoke of my work with kindly, royal condescension. Later I heard that he had remarked of me: 'She's the best man at Harvard'. I record this as an example of the unfortunate results of indiscriminate praise. It was a kindly joke, but Shapley never forgave me. No love was lost between him and Edwin Hubble.

I received one important piece of advice, but I was unable to profit from it. When I talked to Dr Anderson about the problems presented by photometry, he said to me: 'If you are starting out now, I advise you to concentrate on the new photoelectric techniques. They are the coming thing.' This, I regret to say, was the first I had heard of photoelectric photometry. When I got back to Harvard I told Shapley that I should like to undertake work on these lines. But he had other plans. Photographic standards were what he wanted of me.

My visit to Lick Observatory was a comical failure. I wanted to talk to Robert Trumpler about his work on open clusters, but he was away. I wanted to spend the night there, but no room was available. Frances Wright and I had a tent with us (we had used it most of the time in our journey across the country) and we offered to sleep in that. But tents were not permitted on Mount Hamilton. We left the Lick Observatory unceremoniously. Ten years later, when I arrived there with husband and family, we received a royal reception, and Mount Hamilton became a happy memory.

On returning to Harvard I tried to bring some order into existing magnitudes. I undertook to reduce visual magnitudes to a uniform system, attempting to remove the effects of individual color equations, and to bring them into conformity with the International photovisual system. I tried to do the same thing for the photographic magnitudes determined with different telescopes – the problem that had reputedly defeated Miss Leavitt. At Shapley's request I began to establish red and yellow standards photographically. All this, though good discipline, was labor thrown away. As Anderson had predicted, photoelectric standards were the coming thing, and the day of absolute photographic photometry was over.

Many years earlier, when I first began to be interested in open clusters, I realized that the need was not photometry integrated over the photographic and visual ranges, but true spectrophotometry. I

At the cross roads

Fig. 9. Cecilia Payne 1931

projected what I called a system of heterochromatic photometry, based on differential studies of energy distribution. But it was short lived. Photographic methods were soon to be supplanted by photoelectric scans.

I cannot think that I made any lasting contribution to astronomical photometry. My work was like the efforts of the boy in the fairy tale, who always used last time's methods on a new problem, with disappointing results. But photometric work made a lasting contribution to my own development. I learned that one should not use obsolete methods. I know the pitfalls of fundamental photometry, and I recognize good determinations when I see them.

So ended my first decade at Harvard College Observatory. I had written a letter of thanks to Shapley when I had been there a few years: 'You have turned me from a schoolgirl into a scientist, from a child into a woman'. I had seen astrophysics in bud and blossom. The time of harvest was yet to come.

PART III

The dyer's hand subdued

17

Turning point

I had intended this to be an account of the developing ideas of a scientist, sketched on the background of the changing picture of the Universe. The story is one of successes and frustrations, enlightenments and mistakes, loyalties and animosities – all essentially personal. Much as I wish it, the personal cannot be excluded. And when I come to the year 1932 the whole picture grows personal in retrospect.

There has never been a time when my life seemed to have settled into a more steady pace. I think I looked forward to a future devoted solely to the pursuit of scientific work. I must have been blind to what was happening in the greater world. It seems hard now to believe, but as I look back I think that the stock market crash and the depression of 1929 made no impression on me. I had always lived frugally, and my modest salary not only covered my needs, but permitted me to make a monthly contribution to the relief of 'The Unemployed' – several workers who were added to the work force of the Observatory during the financial crisis (and, I fear, sadly underpaid). The future could not have seemed more personally uneventful.

The International Astronomical Union was to meet in Cambridge, with Harvard Observatory as the host, in the summer of 1932. We were all involved in organizing and planning, and it promised to be a busy summer. Dr Shapley was in his element, and we all caught the infection.

Then, early in the summer, Adelaide Ames was drowned as the result of a canoeing accident. For me it was a tragedy so great that I can hardly write of it, and even now I cannot bear to speak of it. In my first year at Harvard we had been inseparable; they used to call us 'the Heavenly Twins'. In later years there had been other friends, but none whom I loved as I loved her. When a blundering acquaintance, who had received a confused account, exclaimed on meeting me: 'Why, I heard you had been drowned!' I replied unthinkingly: 'I only wish it had been I.' I realized that, in comparison to me, she had so much to give, so much to live for. She was beautiful and gay, everyone loved her. She was the only child of adoring parents. She should have been a happy wife, a lovely mother. I was absorbed in my work, shy and unattractive. What was I giving? I made a silent resolve: I would open my heart to the world, I would embrace life and do my part as a human being.

The International Astronomical Union came and went. As a scientific gathering it left no mark on me. But it gave the chance to renew old friendships. In particular I saw Dr Gerasimovič, who arrived, I remember, when the meeting was nearly over, the victim of the deliberate delays of Russian bureaucracy. He remembered our close association in earlier years, and extended an invitation to me to come and visit Pulkova. I had always been rather afraid of him, but now (as later) I realized that his sardonic humor was the cover for a warm heart. I told him I would come.

The following winter and spring were among the saddest times of my life. Resolute attention to work was not enough, and in May of 1933 came a double tragedy: within two days I learned of the death of two dear friends. Bill Waterfield, who had shared so many musical hours with me while he was working at Harvard, was killed in a motorcycle accident in South Africa. And the previous day I had heard of the tragic death of my dear Cambridge friend, Betty Leaf. At Newnham she and I had been inseparable. I can hardly bear to speak of her even now; I have always carried her picture in my wallet – a snapshot made when she and I took a brief bicycle tour together after College was done. The day I had news of her death there was to be a picnic for the Observatory, and I asked Dr Shapley to excuse me, telling him the reason. But he insisted that I should go. 'There's worse to come', he said. For as I realized later, he had received the news of Bill Waterfield's death, and withheld it until the next day so as not to spoil the festivity. I thought – I still think – it was cruel and insensitive. But truly I was too stunned to care.

Adelaide and Betty – all that I was not, beautiful, delicate, beloved – were dead and I was alive. My resolution to open my heart to the world had what I suppose was the inevitable result. I fell in love for the first time – unreasonably, groundlessly, but nonetheless thoroughly (for I am nothing if not thorough). It did not take me long to see that my love was not, never would be, reciprocated, and I fell into a state of despair that seems to me now to have been comically Victorian. I felt that my life had been shattered, that I must make a break with the past. I confided in my good friends, the Boks (without whom I should never have got through that unhappy time), and together we planned that I should travel.

The trip was planned to cover the observatories of Northern Europe. There was a flying visit to England. Here I remember seeing Stratton, he who had told me I could never hope to be an astronomer. He spoke of a visit to Germany: 'They were all saying "Heil Hitler"', he told me, and (so innocent was I of what was happening in the world) I wondered why he spoke so scornfully. I remembered vaguely that at the meeting of the International Astronomical Union I had offered to type something for Professor Kopff and had ignorantly spelled 'National' as 'Nazional'. How horrified he had been! 'That', he had said, 'is a word we do not like.' I was to have a rude awakening later in the summer.

I visited Leyden, the redoubtable Hertzsprung and the resolute Oort, who introduced me to modern Dutch art. Of a lovely still life, by Verster, of eggs in a bowl, he remarked: 'It gives an inner peace'. So did the atmosphere of the Observatory at Leyden. At Copenhagen I visited Elis Strömgren; my chief memory of that visit is of our visit to the Tivoli (Eddington had told me that this was the high spot of the city) and of there being introduced to his son Bengt. I had a feeling that the latter was concerned lest I overdo the potent drinks with which his father plied me! At Lund I visited the Lundmarks, and here I actually did overdo in potations of Swedish Punch (I ought to have remembered Rutherford's anecdote); I learned the lesson that one should know one's limitations. And at Stockholm I was royally welcomed by the Lindblads. Every one of these visits was pleasant, social and superficial. But I was bent on a more serious trip, the visit to Pulkova.

To this end I took a boat from Stockholm to Åbo, my first glimpse of Finland. It seemed, as the boat threaded its way through the Swedish and Finnish Archipelagos, that I was leaving the world I knew and entering another. My linguistic education had not prepared me for Finland. With English, French, German and Italian I had been able to understand and make myself understood everywhere, but in Finland the simplest words, like 'bank' and 'restaurant', seemed quite unrelated to 'pankki' and 'ravintola'. I resolved to learn Finnish when I got back to Harvard, but never succeeded in finding a source of information on that or on the related language of Estonia.

From Helsinki I was to take the short hydroplane flight to Reval. My departure was a scene of comedy. While waiting for the plane at the water's edge, I ventured down the sloping boards toward the sea. It was low tide, and the wooden walk was covered with a treacherous film of green algae. My foot slipped, and I slid very slowly but inexorably into the Baltic. As I stood helpless, waist deep in the water, two airport policemen came rushing to the rescue, waving little white towels, pulled me out and rubbed me down vigorously to dry me off. I was bundled, still dripping, into the plane, and we made our 20-minute flight to Reval. Here I was met by my old friend Ernst Öpik, who tactfully overlooked my waterlogged garments and whirled me off to a tour of that truly remarkable city. Thence we proceeded to Tartu, the site of the historic Fraunhofer Observatory from which the first Struve had gone to Russia.

In Estonia I had a taste, for the first time, of the international tensions that were dividing Europe. It is incredible to me now that I had not realized them before. I said that I was on my way to Russia, and the announcement was met with surprise and resentment. Dr Öpik tried to persuade me that if I visited Russia I should certainly lose most of my property, if worse did not befall me. I insisted that I was going in any case. He insisted that I should not. I remember that we were on a walk through the woods near Tartu, keeping 15 feet apart, and that I was weeping with exasperation. Finally he asked to look at my visa, to see whether it was in order. I handed over my passport, and he pointed out scornfully that the time for which the entry visa had been issued had already expired – implying that it had been done on purpose. As the whole thing was written in Russian I had not been able to discern this for myself.

I think that Dr Öpik believed that he had clinched his argument and persuaded me not to go. But I kept my own counsel. Next day I returned to Reval and went to ask the American Consul for advice. He pointed out to me that the United States had no diplomatic relations with Russia, and that if I 'got into trouble' they would not be able to help me. He, too, advised me not to go. But I insisted. Then he said that he could give me no help in straightening out the visa, as he could not approach the Russian Embassy; however, he offered to give me an introduction to the Foreign Secretary, who could in turn put me in touch with the Russian authorities. So it was arranged. I had the feeling that I was playing a part in a rather grim Gilbertian opera. I passed from one gorgeous office to another, and finally had the satisfaction of seeing my visa updated, with apologies for the 'mistake'. I was on my way. I bought a railroad ticket for Leningrad, and (mindful of Dr Öpik's warning about losing my luggage) deposited everything but a small bag in the Left Luggage Office in Reval.

The railroad trip was rather weird. I was alone on the train after it crossed the Russian border, and I think I was a sore puzzle to the two GPU men who examined me as we entered. They insisted on counting personally all the money I had with me (which caused a minor embarrassment, as I was carrying it all in a money belt underneath my clothes). They also went, piece by piece, through my luggage. The only item over which they hesitated for a long time was a postcard reproduction of Blake's painting *The Ancient of Days*. This obviously represented God, and though they seemed doubtful about admitting Him to the Soviet Union, they finally returned the Almighty to me.

Arrived in Leningrad, I knew a moment of panic. A bleak impersonal station, no porters, no comprehensible signs. And then, blessed sight, Gerasimovič appeared. He was embarrassed, he had come with a pickup truck and a driver; it was not legal for three people to ride in the front seat. So I rode out to Pulkova in state on the rear floor of the truck.

I spent two weeks at Pulkova, and felt I had experienced a lifetime. The atmosphere of tension never lifted. It was not only the drab and squalid living conditions of the man who was Director of one of the great Observatories. It was not only the scarcity of food – for food was severely rationed and they shared their rations with me. I had brought them some coffee, and they gave a party to celebrate it – nobody there had tasted coffee for several years. One day there was a special treat for supper – carrots, and my host confessed that he had stolen them from a neighboring garden. Small wonder that the food nearly choked me – unappetizing as it was, I was taking it from their plates. Everyone was afraid – afraid to talk lest they should be overheard. One of the young women – she has long been dead now – led me to the middle of a wide field and begged me in a whisper to help her to go abroad – 'I would wash dishes', she said, 'I would do *anything* to get away from here'. And what could I do? What could I possibly have done? I was appalled.

The situation had its laughable side too. I was shown over the Observatory with its historic instruments, including the Great Telescope that had been the twin of the Harvard 15-inch, then the world's largest telescopes. When we came to the Bredichin Astrograph, they warned me to be careful – there was a bee's nest behind the shutter. How long, I asked, had it been there? *Seven years!*

Foremost among memories of that visit is my meeting with Belopolsky, the 89-year-old astrophysicist who had made history before the turn of the century. He was still active, still interested in everything. 'Schon weiss ich viel, doch möcht ich Alles wissen' he quoted to me from *Faust*, and it was true. He told me a tale of his Directorship of the Observatory, a few years earlier. There was no wood for the fires and the Observatory was freezing. He took a cart and went out, returning with a load of wood. Then, he said, he presented himself before the local authorities. 'I am an old man', he told them, 'and the Director of a great Observatory. And in order to keep it warm I am compelled to go out and steal my neighbor's fences.' The next day, he said, a load of wood was delivered to the Observatory. I can well believe it of him. He was without fear. I shall not forget my last sight of him. I went over to say goodbye, and found him covered with flour. Insects had been discovered in the flour bins, and he was himself putting all the flour through a sieve.

As we drove back to Leningrad, I saw masses of troops and army

Turning point

planes. I asked what it was. 'We are preparing', said Gerasimovič gloomily, 'against our Enemy.' 'What enemy?' I asked. 'Germany', he replied.

The two weeks I spent at Pulkova would never be effaced. I felt that my personal griefs had been obliterated by the human tragedy. We all sensed, I think, that we should not meet again. When I left, Gerasimovič made me a present of an embroidered tablecloth. I shall never forget the sadness in his face as he said: 'It is a custom with us, when a friend is going on a journey, to give him a tablecloth'. And he died a few years later, a victim of Stalinist persecution. He knew what was in store for him. He spoke very little of the situation, but once he said: 'When I saw what was becoming of my beloved country, I wanted to kill myself'.

When I was on the train for Berlin I felt as though I had held my breath for two weeks. As we left Russia I lost the sense of oppression, and when I changed trains in Reval I succumbed to hunger, and bought a small roast chicken, the first meat I had seen in two weeks. It was an ill-fated purchase, for it produced a violent bout of food-poisoning. On the station platform in Riga I thought I was going to die, and my one desire was to get on the Berlin train – I had had enough of countries whose language was a complete mystery, and was in terror of a Latvian hospital.

I did indeed reach Berlin, shaken but alive. I could scarcely eat, and was in two minds about going on to Göttingen to the meeting of the Astronomische Gesellschaft. But some good angel prompted me to take courage, and soon I was on the train again. It was good to be in Germany, where I could understand and be understood, but it did not take me long to sense that all was not well here either. A horrible, leering old man got into conversation with me on the train, ranting incomprehensibly about politics. When I took refuge in my little notebook he asked: 'Ist das Ihr Tagebuch?' and actually put out his hand for it. This was not Russia, but it was almost as alarming.

Then I was in Göttingen, at the meeting of the Astronomische Gesellschaft. Everyone was kind, but there was a feeling of tension and distance. I knew none of the German astronomers, and though Eddington was there, his place was among the great. I took my seat shyly in the big auditorium, and someone brought me my mail,

Fig. 10. Sergei Gaposchkin, Cecilia Payne-Gaposchkin 1934

pronouncing my name as he did so. A young man sitting near by looked up in surprise. 'Sind Sie Miss Payne?' he asked. I admitted as much. He introduced himself. His name was Gaposchkin, and he had come, he said, in the hope of seeing me. I do not think we spoke much, but he put into my hand a statement of his history, and asked me to read it. (He had expected, as I learned afterwards, that Miss Payne would be a little old lady, and was surprised to find her no older than himself.)

When I got back to my quarters I read the history he had given me, and I learned that here was one who had resolved, as I had resolved, to be an astronomer, and against what terrible odds he had achieved what had come so easily to me. Of course I knew I must help him to escape the last of the many disasters that had overtaken him, Nazi persecution, and to establish himself in a new world. I have not spent many sleepless nights, but that one was sleepless. Perhaps this, I thought, is my one chance to do something for someone who needs and deserves it. When I saw him the next day I told him that I could make no promises, but I would do what I could.

Turning point

When I tried to talk to the German astronomers about the subject, they were anxious and evasive – always looking over their shoulders lest someone should overhear. I had seen the gesture in Russia too. Yes, they could recommend him; yes, he was a good astronomer. But it was impossible for him to stay . . . I began to understand the political climate. He had been born in Russia; that was enough. (Who is your enemy? I had asked Gerasimovič; the picture fitted together.)

This chapter is long enough. Only when I was back in the United States did I feel I could breathe freely. And here it was possible to act freely too. I had never tried to exert any influence before, but I tried it now. A place was found at Harvard; I went to Washington to expedite the granting of a visa to a stateless man. Between August and November the thing was done, and Sergei Gaposchkin set foot in the New World.

What was the turning point? The death of dear friends, the pangs of unreciprocated love, the spiritual climate of Russia, of Germany? Perhaps it was all ordained from the beginning. It led to the uniting of two lives, the flowing of two rivers, bound for the same goal, into one channel. In March 1934 I became Cecilia Payne Gaposchkin.

18

Prolegomena to variable stars

It was the beginning of a new life. We set off on a honeymoon trip across the country, retracing the steps I had taken four years before towards the great observatories of the West. In the interval I had settled down to the serious business of life. From now on the effort must be systematic and coordinated. I felt I was no longer a freelance, but a member of a team.

My Husband had produced a doctoral thesis on eclipsing stars, truly remarkable for its thoroughness and scope. It had been the right moment for bringing together the enormous body of existing data on this group of variable stars. The time had come to consider what was known about them, and what was still needed for their interpretation.

Some years earlier, in the course of my survey of the stars of high luminosity, I had had occasion to survey the knowledge of intrinsic variables, and had made a beginning in my own understanding of their relationships. My early impression, that all intrinsic variables are luminous, had long given way to the recognition that they occur at many levels – the highly luminous supernovae and novae, the luminous Cepheids, the less luminous Mira stars and RR Lyrae stars. A synthesis of our knowledge of the low luminosity of novae and U Geminorum stars at minimum was still to come, but variable stars were emerging as having a great variety of properties. It began to seem possible to classify them in a coherent scheme. No longer could they be regarded as 'pathological stars'. They clearly occupied definite domains in the range of stellar properties. Nor were they the province of the amateur astronomer, as they had seemed to be a few decades earlier. Miss Leavitt's discovery of the period–luminosity relation in the Magellanic clouds, and Shapley's work on the globular clusters, had put an end to that. Variable stars were serious business in astronomy now. In the hands of Henrietta Swope, the Harvard programs on variable stars in the Milky Way were bearing fruit in enormous quantities.

It seemed to us that the time had come to put together what was known about variable stars. Together we undertook a survey of the subject, my Husband covering the eclipsing variables, and my own domain being all the rest. Our goal was the arrangement of variable stars in a coherent scheme, with an attempt to cover the known species, and *Variable stars* was the outcome of our efforts.

The time was ripe for such a survey. Twenty years earlier the data would have been too sparse. Now, more than 30 years later, they would be unmanageable, and the subject calls for specialized treatments. In 1935 the picture could be sketched in broad outline, and in its main features it is still valid. It furnished us with a game plan for further exploration, a project for new researches.

Not only was the time ripe, but the place was right for a synthesis. The Harvard plate collection lay ready to our hands. In those days the Harvard Observatory was the scene of a weekly meeting, called by Shapley the 'Hollow Square' because we were ranged around tables placed side by side in square formation. It was the occasion of free-for-all discussion of timely scientific topics. One afternoon Shapley was expatiating on the riches of the collection of photographs when a thought struck me. 'Yes', I said, 'but are we making full use of it?' He challenged me to make good my criticism, and at that moment an idea was born. Why not use the Harvard plates to get all possible information about all the known variable stars? I had already tried my prentice hand on a couple of hundred Mira stars (and alleged Mira stars that turned out to be something else). It was an obvious plan, an ambitious plan, but feasible. We undertook to present an outline for the project.

Of course we could not investigate *all* known variable stars. I had not forgotten my first essay in systematic photometry. The plan had been to measure photographic magnitudes on extrafocal plates over the whole sky down to a given limit. In my youthful ambition I projected a program for all stars down to the twelfth photographic magnitude. I was quickly deflated by Shapley. Did I know how many stars would be included? I did not. How many times must each star be measured, and how long would each measurement take? I had not thought of that. For one person, working full time (which was all I could count on) it added up to about 100 years, and this did not count observing time or false starts. As it turned out I only completed the north polar cap, down to 75°, to a seventh magnitude limit. Unwilling to be caught the same way a second time, I surveyed the size of the problem and decided that a reasonably tractable program would include all known variable stars that had been recorded as brighter than the tenth magnitude at some time. The type of magnitude was not specified, and, existing magnitudes being what they were, a number of much fainter stars were inadvertently included. At the current stage of variable star discovery, the total number was about 2000.

Clouds were gathering on the international scene, and Dr Richard Prager, renowned for his catalogue of variable stars, had left Germany and come to work at Harvard. His reaction to our plan was that it would take us 90 years to complete. It is pleasant to relate that we completed it in five. Funds were provided, and half a dozen people were hired to do the measuring. Even so, the task was a formidable one. A bright star in the northern hemisphere could be studied on at least 5000 plates, and (except for stars with brief outbursts) most could be found on at least several hundred. The secret of the efficiency and success with which the program was carried out was the method of estimation devised by Sergei Gaposchkin. Everything was done systematically and objectively; the assistants made the estimates and the Gaposchkins did the reductions.

The material on which this survey of the brighter variable stars was carried out had its strengths and its weaknesses. The strength lay in the large number of available photographs and in the long time base; the first plates dated from 1889. One weakness lay in the fact that the photographs were part of a routine program, not geared to particular problems. The brightest stars were too bright to be accurately studied, for the exposures had been aimed at maximum coverage. The lengths of exposure set a limit too, on the study of very rapid variations or very short periods. The plates were taken with a variety of instruments, and color equation introduced problems, especially for stars of extreme color. Most serious of all, the establishment of magnitude scales presented great difficulties, and the most that could be claimed for the final magnitudes was that they were internally consistent.

The fruit of our five years' labor, then, was primarily the accurate determination of periods, and of changes of period. The forms of the light curves were well enough determined to be significant, and the magnitudes were at best approximate. We were able to clarify the type of variation for many stars and to classify many others for the first time.

I can mention several interesting surprises. When we outlined the program, the question arose as to whether well-observed novae were worth re-examining. Dr Prager expressed the opinion that it was not worthwhile. We decided nonetheless to include them. In consequence the nova U Scorpii, which erupted in 1866, was found to have recurred in 1906 and in 1936, a new discovery of a recurrent nova. The images had been waiting for several decades to be found. Nor is this an isolated case: WX Ceti, long recorded as a nova, was recently observed to recur by Sergei Gaposchkin.

A surprise of a different kind was presented by RY Scuti. One afternoon we decided to 'look up some interesting star', and picked this one out of the list, published by Paul Merrill, of 'spectra intermediate between those of stars and nebulae'. Rather expecting a nova, I remarked as we sat down to work, Sergei to measure and I to record: 'We can be sure, at least, that it is not an eclipsing star'. Half an hour later we knew that it *is* an eclipsing star, and one of the most remarkable on record. Before the day was over we knew that it is a counterpart of Beta Lyrae; it is greatly reddened and greatly obscured, surrounded by a bright nebula.

Bright stars, as I have said, are hard to study on routine exposures, but some of the most spectacular eclipsing systems have been found among them on the Harvard plates. It takes a keen eye to estimate variations for an overexposed image, but it was Sergei who first discovered the eclipsing nature of VV Cephei, UW Canis Majoris, V453 Scorpii, Upsilon Sagittarii, CQ Cephei, and V444 Cygni, each of which has made history in its own way. The details of their variations call for the refinements of photoelectric photometry, but without the Harvard plates they might still be undiscovered. I believe that the possibilities of the Harvard collection are still unexploited. It would even be worthwhile to bring up to date the stars brighter than the tenth magnitude that have been discovered since our survey was completed 30 years ago.

The book on variable stars that was our first collaborative work made no pretense to finality. It could truly have been entitled 'Prolegomena', indeed that was the title we originally wished to give it. The study of variable stars on plates of the Harvard collection is of the same nature, and can claim to be no more. Here we can pick out star after star that is worthy of intensive study. This is no place for final and definitive work; it is what Pickering intended it to be, the domain of the pioneer.

19

International problems

The European scene was darkening. When we were invited to take part in a Conference on Novae and White Dwarfs in Paris in July 1939 we were in doubt whether it would be wise to go. But we wanted to bring our two small children to meet their relatives in England. Finally we decided to take the chance. It was to be our last sight of Europe before the catastrophe.

It was the first unified conference on novae. Eddington and Stratton were there from England; Russell, Chandrasekhar, Baade and Kuiper and ourselves from the United States; Beals from Canada; Lundmark from Sweden; Bengt Strömgren from Denmark; Chalonge, Mineur and Barbier from France. It was a happy meeting; discussion was free and uninhibited. As I look back, I see that we knew little about novae in 1939, and that little was confined to the phenomena of the outburst. Whoever planned the conference had a prophetic instinct when he combined novae and white dwarfs as the subject, for the close link between them was still unsuspected.

At the final dinner I sat between Eddington and Frédéric Joliot. It is a poignant memory that Eddington said to me: 'Do you think your Husband would mind if I took you in to dinner?' In those words he linked the dreams of the far past with present realities. It was the last time I ever saw the greatest man I have been privileged to know.

Frédéric Joliot I met for the first time. I found him completely fascinating. He spoke of his current research, telling me that he and his wife, Irène Curie-Joliot, were observing chain reactions. How often I remembered that conversation in the years that followed, and trembled for the Joliots.

Fig. 11. Paris Conference 1939. (From left, back row) C. S. Beals, B. Edlén, P. Swings, B. Stromgren, S. Chandrasekhar, W. Baade, (middle row) G. P. Kuiper, K. Lundmark, (front row) F. J. M. Stratton, C. Payne-Gaposchkin, H. N. Russell, A. Schaler (conference organizer), A. S. Eddington, Sergei Gaposchkin (Photo courtesy of S. Chandrasekhar)

Paris was still gay; we were not conscious of the approaching storm. We watched the colorful parades of Bastille Day, French and Belgian troops marching by the Arc de Triomphe. Then we returned to England, and I relived my youth as we sported with our children at a seaside resort.

The storm broke on the day we returned to London. I shall never forget the panic scene in Whitehall, the crush of cars, people rushing to and fro between the Government offices. We had passages on the *Normandie*, in what proved to be her last voyage. War was declared when we had been two days at sea, and we made an eerie, blacked-out passage to New York.

We returned to Harvard and to work. The joy had gone out of life, and many times I wished myself in England. Routine work was a blessing. Variable stars and the responsibilities of a household filled our days. As the United States came closer to the brink of war we considered our obligations, and when war came we made our own attempt to fulfil them.

Our first step was to purchase and equip a poultry farm in central Massachusetts. We had a double purpose: to contribute to the food supply, and to give work to a refugee family. The first object we did fulfil, but we found no refugees who were interested in farming. But the farm raised pigs; we bought a cow whom we called Victoria, and a sheep whom (inevitably) we named Albert. We saved fuel by investing in a horse and buggy – the former was named Stanley (from a fancied resemblance to the Prime Minister), and for a time practised the simple life.

The task of carrying a hundred dozen eggs, innumerable broilers and (later) several hundred turkeys to market occupied our spare hours, but it was not enough to satisfy us. As the European situation deepened and darkened, we felt that people knew too little of its background. So we projected a Forum for International Problems.

Shapley gave willing support to our plans. He allowed us to use the Observatory to hold a weekly meeting, and for speakers we drew on the rich international background of the University and the local community. Our motto was: 'Good will is not enough; there is no substitute for knowledge.' There was no lack of subjects: England, France, Italy, Spain, Germany, Russia, India, Hungary, Israel ... spokesmen were easy to find. We always insisted that as many sides of a problem as possible should be presented, so the meetings were debates rather than single presentations.

With an assurance born of desperate good will I assumed the duty of Chairman. It was no sinecure. The speakers often urged their arguments with intemperate zeal, for feelings ran very high in those days. It was hard for a chairman to be dispassionate, hard not to take sides when I longed to be a partisan. There were times when I feared physical violence on the platform. But for several years we persevered in our struggle for enlightenment.

There was no difficulty about attracting an audience. Every meeting was packed with eager listeners; there were lively discussions and arguments from the floor as well as between the speakers. In fact the discussion finally became too lively. Each side in the debate was supported by a claque, who cheered or booed as occasion required. We felt that our Forum had fulfilled a purpose and outlived its usefulness, and left the further pursuit of clarification to others.

It was an episode that taught me much, not only about international problems but about human relationships and communications. We both look back on it with satisfaction. But in spite of my strenuous attempts to keep the direction impersonal and dispassionate, the Forum earned me the reputation of being a dangerous radical. I still resent this as a grave injustice, but it is not infrequently the fate of one who tries to work in the cause of truth and justice. I think of the words of one of our speakers, Professor Rankine of Tufts, a saintly man who had lived in India and had known Gandhi: 'I'm more extreme than the extremists, more radical than the radicals. What am I? Just a Christian.'

20

End of an era

When I came to Harvard Observatory in 1923, Harlow Shapley had been the director little more than a year. The death of E. C. Pickering had left a vacancy that was hard to fill. Henry Norris Russell told me many years later that the position had been offered to him, and that the decision to refuse had been a hard one. What would have been the effects on the development of astronomy if he had accepted? Instead of doing that he recommended Shapley.

It had been Pickering's wish, I have been told, that Solon Bailey should succeed him. The University hesitated to appoint the young middle-westerner immediately, and invited him instead for what was essentially a trial period. It came to Bailey's ears that the Powers were uncertain whether Shapley was suitable for the position, and that his own appointment was being considered. He went (so I have heard) directly to the President of the University and told him that it would be folly not to appoint this brilliant young man. And as soon as the appointment was decided upon, Bailey left Cambridge for the Observatory's southern station in Peru, leaving the new Director a free field. It was a typical action of this meticulous, dedicated scientist to put the good of the Observatory before his own advancement.

In an earlier chapter I have given an account of the institution that Pickering left for his successor. At the time it was the greatest storehouse of astronomical data in the world. It had begun as a one-man operation, and its undertakings had remained so; although Miss Leavitt, Miss Cannon and Miss Maury had emerged as independent scientists, their efforts had been channeled according to the plans of the Director.

I have spoken of the way in which Shapley transformed what he found. Adelaide Ames and I had a joke about it. 'It used to be a dear little Observatory and he is going to make it a Great Institution', we said. He began to appoint young scientists who did independent work; he expanded the publications, and they began to appear under other names than that of the Director. He exploited new techniques. He organized a series of talks on astronomy for radio broadcast – a great departure for that time.

The Observatory was treated as a big family. Social events in the Director's residence embraced the whole staff from top to bottom, and invitations were treated as the commands of royalty, under no circumstances to be refused. It used to be said that the *sine qua non* of a position at the Observatory was an ability to play bridge and ping-pong. In fact, Shapley said he could assess the suitability of an aspiring staff member by his demeanor at the latter game. Sometimes the parties were enlivened by dancing, or by elaborate charades. I can still see Shapley impersonating Napoleon (he prided himself on the resemblance), and Henry Norris Russell, swathed in a sheet, acting out the Death of Archimedes.

An even more ambitious caper was the presentation of the 'Observatory Pinafore' for a meeting of the American Astronomical Society – a skit on the Gilbert and Sullivan operetta, satirizing the doings of the Observatory at the time of the original production. The whole staff was pressed into service, and poked faded fun at Pickering's early essays in photometry, with vexatious prisms that showed polarization. I was promoted to the chief woman's part, that of the heroine Josephine. Alas, it is a soprano part, and I was a contralto. No matter, it had to be done, with the result that I have never been able to sing since.

Shapley came to Harvard on the heels of his long series of papers on globular clusters – work that made scientific history and placed him among the astronomical greats when still in his thirties. His first interest in his new position was the size and structure of the Galaxy, and, as I have already mentioned, he began an ambitious program of the discovery of variable stars in low and high galactic latitudes. The results were surprisingly fruitful – even more so, perhaps, than the planner anticipated – and the program was only partially completed.

End of an era

The data proved to be harder to interpret than had originally been expected, for Shapley was reluctantly compelled to concede that interstellar absorption played a large part, with results difficult to assess, in the apparent distributions. This was particularly serious in the direction of the galactic center, where much of the first effort was concentrated.

It is not easy now to recreate the picture of the Cosmos (as Shapley was inclined to call it) on which his observational programs were based in his first days at Harvard. Impressed as he was by the great size of the Galaxy, his emphasis was on dimensions rather than structure. And from the first the Magellanic Clouds were an integral part of the picture. Clearly they were distinct entities. I had not forgotten Hertzsprung's evaluation of their distances, which was given at the meeting of the Royal Astronomical Society where my first small paper saw the light. In the Clouds alone was the period–luminosity relation directly observable.

It seemed plausible then – and for many years thereafter – that the period–luminosity relation for the Magellanic Cepheids was coincident with that for the Cepheids in globular clusters. So the distances of the Clouds were thought to be much less than they are now known to be (the uncertainties of the magnitude scale conspired to exaggerate the effect), and they might well appear to be outlying clouds of our own system, comparable in many ways to the Cygnus and Scutum Clouds. Only with hindsight can we trace the transition by which the Magellanic Clouds emerged as isolated stellar systems.

The nature of the external galaxies had not been established when Shapley took the helm at Harvard. He had defended their local status in the famous debate with Curtis. 'I believed in van Maanen's results (on the rotation of spirals)', I heard him say ruefully many years later. 'After all, he was my *friend*.'

Hubble's discovery of the Cepheids and the period–luminosity relation in the Andromeda Galaxy was a bitter pill for him to swallow. I was in his office when Hubble's letter came, and he held it out to me: 'Here is the letter that has destroyed my universe', he said. Small wonder that his private response and his public comments were less than enthusiastic. But his actions spoke louder than his words. From that day he planned a gigantic assault on the problem of galaxies, work that would occupy him for the rest of his scientific life. The emphasis of Harvard astronomy shifted abruptly from the local area to the distant systems.

During the nearly 40 years of his directorate, Shapley attempted to broaden the Observatory's facilities. Observing was moved from Cambridge to a country site, and the southern station was transferred from Peru to South Africa, though the ambitious plans were never matched by adequate funds.

Pickering had maintained a research institution. The new regime projected a Department of Astronomy with an active graduate school. I had been the first to receive a PhD for work carried out at Harvard Observatory. Students were soon attracted in large numbers, and for several decades the Harvard graduates filled positions in many observatories, not in the United States only, but throughout the world. Visitors from every country passed through to lecture and to discuss. There was not a great name in contemporary astronomy that did not visit us. The 1930s were a heady time at the Observatory, one of Shapley's chief contributions to the growth of astronomy.

As the years passed, his interests broadened rather than narrowed. 'I never go to Church', he once remarked to me, 'I'm too religious for that.' And, though I suppose he would have described himself as an agnostic, he was sympathetic with religious movements, and a frequent sharer of their councils. His worldwide contact with scientific men awakened him to the growth of international tensions, and he threw himself with energy into the arena. He publicly espoused liberal causes. These excursions into politics inevitably earned him the reputation of being a dangerous radical. This reputation, for which he suffered much, was in fact quite undeserved. I knew him well enough to realize that he was no radical: he was a conservative at heart, and ran his own observatory like a benevolent dictator. But he was an implacable foe of injustice and fraud. The hostility that he incurred during the internationally turbulent years cast a shadow that saddened and weakened his final years as director of the Observatory.

In 1959 Shapley's term of office came to an end. Stellar and galactic astronomy had passed through a golden age, and he had played a not inconsiderable part in it. He had indeed built up a Great Institution.

It was a hard act to follow. He was reluctant to leave the stage, and like Pickering he would have liked to perpetuate his regime and select his successor. But it was not to be.

Many years before, when he was considered for another appointment, he said to me: 'We must all ask ourselves, have we picked our successors?' I was so young and unrealistic that I replied: 'I don't mind your leaving, if I can succeed you'. Had I been a man, perhaps he would indeed have wanted to hand over the throne to me. But it would have been a great mistake. I should have tried to maintain the pattern that he had established. But just as he transformed the institution that Pickering left into a research group and an active teaching department, the institution that he had built must respond to the changing times and the changing face of science.

After an interval of uncertainty and tension, Donald Menzel succeeded Shapley as Director. The changes that followed can be illustrated by the physical transformation of the observatory. It was not long before the Smithsonian Astrophysical Observatory, under the directorship of Fred Whipple, moved to Observatory Hill, and the two institutions combined their resources. Spacious new buildings have replaced the classical wooden structures that I remember; the only survivor of the old days is the 'Brick Building', originally designed to hold the photographic collection, now converted into office space. The enormous, cumbrous Director's residence, once the scene of the innocent gaiety of the observatory family, is a thing of the past, and a modern office building occupies that once romantic spot.

The new regime brought changes to me also. As I relate elsewhere, it was Donald Menzel who saw to it that I became a Professor, and I was promoted to the Chairmanship of the Astronomy Department, which proved to be anything but a sinecure.

During the past years the emphasis in astronomy has changed dramatically, and we have changed with it. Satellites, the expanding baseline in wavelength, and high-energy astrophysics have opened up vistas that were beyond imagination 50 years ago. Leo Goldberg succeeded Donald Menzel – the first Director of the Observatory who had been a student there. And today we have a synthesis of the two Observatories in the Center for Astrophysics, under the single directorship of George Field. The 'dear little Observatory' that Adelaide Ames and I joked about more than 50 years ago seems like a dream.

So does the small stellar world in which we moved in those days. The Universe has expanded beyond belief (and that in a non-relativistic sense). And the Observatory that does not expand with it is a dead institution, unworthy of the purposes for which it was designed.

21

Retrospect

This has been an open-ended account of the life of a scientist, broad at the beginning and tapering to a point. I see a good reason for this. At every stage one measures one's experiences by what has gone before. At four years old, a year is a quarter of one's life, perhaps half of remembered life. At 80 it is little more than a hundredth. Perhaps this makes early experience seem more vivid and more significant than those of later years, when wonder has been blunted by repetition.

Having lavished many pages on a few short early years (which seemed long at the time), I now pack more than 10 later years (which have seemed very short) into a few paragraphs. They are a record of steady, systematic garnering and arrangement of facts. I have not been one who fashioned new theories, as I once dreamed of doing; if I have made a contribution, it has been by collecting, turning over in my hands, comparing and classifying the data of astronomy.

For many years the Gaposchkins looked longingly at the Magellanic Clouds, the greatest storehouse of variable stars accessible to us. But the Clouds were Shapley's jealously guarded preserve. He drew on them for representative facts, for data that bore on their sizes and distances, and, most important of all, he maintained a steady flow of photographs over the years. After his retirement we were free at last to put a long-projected plan into operation.

It was in 1904 that Miss Leavitt published her lists of variable stars discovered in the two Clouds. Even those that had been studied since then had not exhausted the available material, and many had not been investigated at all. Meanwhile a large number of additional variables had been discovered on the Harvard plates. For decades, Harvard Observatory had enjoyed a monopoly of the Magellanic

Fig. 12. Cecilia Payne-Gaposchkin 1948

Clouds, but by 1950 other observatories were in a position to study them, and the long-felt want of photoelectrically based standards was being supplied by Gascoigne and Arp.

Our plan was to obtain observations of all the known variable stars in both Clouds on all available photographs, to determine their types and periods, and to relate them to the structure of the two systems. It was an enormous program, involving nearly 2 000 000 observations, and we completed it in five years. The experience gained in our earlier assault on the brighter stars of the Galaxy, and the methods that Sergei had put into such successful operation then, stood us in good stead.

The results embrace between 3000 and 4000 variable stars, about equal numbers in the two Clouds. Most of them are Cepheids, far more than the number of Cepheids known in our own galactic system. Eclipsing stars are potentially an equally important group, for (like all eclipsing stars) they contain the possibility of determining absolute dimensions and masses, and thus indirectly of establishing absolute luminosities and solving the still uncertain problems of the distance scale.

Miss Leavitt's discoveries, and the later Harvard searches, had not exhausted the survey of variable stars in the Magellanic Clouds. We discovered several hundred more, inadvertently, in the course of measuring the known ones. A brief survey with the blink microscope suggests that hundreds, perhaps thousands, still remain undiscovered, particularly in the Large Cloud.

And we have merely scratched the surface of these two nearest stellar systems, skimmed the froth from two immense cauldrons of developing stars. If our notion of their distances is correct, RR Lyrae and Mira stars will be accessible when observations are pushed to the twentieth and twenty-first magnitudes. Stellar populations anything like our own will then reveal myriads of variable stars; and if this expectation is belied, we shall know either that the distances were incorrect or that the populations are dissimilar. There are indeed RR Lyrae stars known in the globular clusters associated with the Clouds; what is the role of the vast numbers of mini-clusters to which Sergei has drawn attention? I can hardly wait to learn the answer.

In an era of new techniques, fresh vistas open up faster than they can be surveyed. I have spent the past few years in contemplating, and turning over in my hands and in my mind, the surveys of stars whose ultraviolet light has been recorded by satellite, and (at the other end of the scale) those that are being observed in the infrared and the near infrared. Perhaps the latter will furnish the clues to understanding the periodic red stars, the Mira stars that seem the most enigmatic of the regular intrinsic variables.

Among the success stories of variable star astronomy during the past 50 years has been the growth of our understanding of the novae and other cataclysmic variables. When I entered astronomy, they were known from the variety of their light variations and the enigmatic changes of their spectra. I have seen the subject develop, first by analysis of the physics of the explosion, then by observation and interpretation of the underlying star. Today the novae, the dwarf novae and the potential novae are embraced in a unified picture of a close double system consisting of a camouflaged white dwarf and a low-temperature star that fills the Lagrangian surface and from which matter flows to the blue component. The novae have been one of my continuing interests, and I foresee that I shall spend the years ahead in turning over their variations in my hands and in my mind. The reconstruction of the geometry of the explosions is becoming a possibility, and an understanding of the springs of their behavior seems within reach.

As I look back on 50 years of contemplating the variety of stars, I sense that some sort of order has emerged from the apparent chaos. We think today of a star not only as a physical object whose condition can be studied, but as a developing entity whose past and future can be predicted. Observation and theory have joined hands in an understanding of the successive stages of a star's career. Stellar development can be studied on the background of star clusters, groups similar in origin and history. It is a far cry from the semi-empirical picture of star clusters that emerged from Shapley's early studies at Mount Wilson, and was described by him in his early monograph *Star clusters*. The work I did for him in editing that book gave me a taste of the enormous possibilities of the subject, and an interest that has never been outgrown. If I were starting out today, this would be my point of departure.

PART IV

Reflections

22

On being a woman

A woman knows the frustration of belonging to a minority group. We may not actually be a minority, but we are certainly disadvantaged. Early experience had taught me that my brother was valued above me. His education dictated the family moves. He must go to Oxford at all costs. If I wanted to go to Cambridge I must manage it for myself. Early I learned the lesson that a man could choose a profession, but a girl must 'learn to support herself'. Presumably this would be until she found a husband. But it was early impressed upon me that I could scarcely hope to do that, as I had 'no money of my own'. Such was the Victorian social code in which I grew up.

In my case the real obstacle to marriage was that I met no men at all. There was an unwritten law in our house that if my brother should bring any of his friends home, his sisters must make themselves scarce. This was part of the social code of the contemporary public school boy – another aspect of sex discrimination.

Once or twice I was asked to a dance, given for some school friend as a 'coming-out party'. This was a concentrated agony. I did not know how to dance. My clothes, too, were an embarrassment, for they were hand-me-downs from the daughter of a wealthy friend. I still remember my horror when I learned that one of my dancing partners knew her, and thought with crimson shame that he probably recognized the dress I was wearing. Even when I fell back on conversation it was a disaster. A friend of my brother, whom I had tried thus to entertain, remarked to him later: 'Fancy! A girl who *reads Plato for pleasure*!'. I simply did not know how to behave at a dance.

Matters did not improve when I went to Cambridge. Women were segregated in the lecture room. Even in the laboratory they were paired off if possible, and (did I imagine it?) treated as second-class students. It might have been different if I had been gay and attractive and had worn pretty clothes. But I was dowdy and studious, comically serious and agonizingly shy. The Demonstrator in the Advanced Physics Laboratory told someone (who kindly repeated it to me) that I was 'slow'. It did not occur to me to protest. Ignorant and uncouth I might be, but not *slow*! I decided to pay no more attention to anything Henry Thirkill said: he was simply not noticing. Unluckily for me, he was one of the final Examiners in the Tripos, and I believed him responsible for placing me in the second class. I heard through the grapevine that the other Examiner, William Bragg whom I adored, had wished to place me higher. Henry Thirkill had put my back up; had I produced the same effect on him?

The attitude to women that oppressed my childhood and youth was typical in England at the time. Fifty years have not mended matters much. Although my work was well known by the time I was 30, I am sure that I stood not the slightest chance of obtaining a position in England between the time I went to Harvard to the time I retired in 1965. And how I would have jumped at the chance! But though I had gone to the Right University, I had read the wrong subject. One could not have become an astronomer in England without having obtained a First Class in the Mathematical Tripos. And, of course, I was a woman. The Royal Observatory was administered by the Admiralty. The redoubtable H. H. Turner recorded that when a candidate for the position of Chief Assistant at Greenwich was asked what qualifications he had had for the job, he replied: 'Among other things I had to climb a rope'. I should have failed the test; rope-climbing has never been my strong point. A restriction to the male sex no longer dominates the Royal Observatory, but something else still has a stranglehold on Astronomy in England.

We manage things better in the United States. Even 50 years ago a woman might do astronomical research and even make a name by publication. She might hold a position – without a title and ill-paid, it is true – and she could meet on equal terms with any astronomer in the world. In my early days at Harvard, everyone who was anybody (and many more besides, who were going to be somebody in the future) came through, and argued, and fraternized. Those were glorious days. We got to know Lundmark, Milne, and Unsöld, Hund, Carathéodory and ten Bruggencate. How we argued, how we walked about the streets and sat talking in restaurants until the manager turned off the lights in despair! We met as equals; nobody condescended to me on account of sex or youth. Nobody ever thought of flirting. We were scientists, we were scholars (neither of these words has a gender). In that heady atmosphere a woman did not degenerate into the abominable stereotype of the *Femme savante*, that combination of conscious erudition and affected coyness that suggests 'It's really not *womanly* to know as much as I do'. How different from the attitude described by one of my English friends: 'With my education, I never could expect to marry'. Yes, we do things better here.

There are those – and I am one of them – who rebel at having to deal with an intermediary. They want to go to the fountain-head. Someone who knows me well says that science, to me, has been a religious experience. He is probably right. If my religious passion had been turned towards the Catholic Church I should have wanted to be a priest. I am sure that I should never have settled for being a nun. If it had been directed towards medicine, I should have wanted to be a surgeon; nothing would have persuaded me to be content to be a nurse. As I look over the world of science, I picture most of the many women who are working in that field today in the role of nuns and nurses. They are not allowed – they are not supposed to be fit – to be in direct touch with the fountain-head, whether you call it God or the Universe. (But even as I write, this situation is changing.) Here I have had no cause for complaint. I have always been in direct touch with the fountain-head. No other mortal has made my intellectual decisions for me. I may have been underpaid, I may have occupied subordinate positions for many years, but my source of inspiration has always been direct.

I spent many years at Harvard, research and writing my main interests, with an undercurrent of editing that gradually took more and more of my time, and incidentally taught me much about the craft of writing. I had no official status, as little as that of the students who provided the 'girl-hours' in which Shapley counted his research expenditures. I was paid so little that I was ashamed to admit it to my relations in England. They thought I was coining money in a land

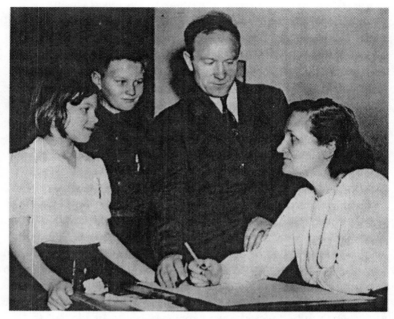

Fig. 13. (From left) Katherine, Edward, Sergei Gaposchkin, Cecilia Payne-Gaposchkin, in her office 1946

of millionaires. But I had the run of the Harvard plates, I could use the Harvard telescopes (a dubious boon, this, in the climate of Cambridge), and I had the library at my fingertips.

Then came the time when Shapley organized the Department of Astronomy, and began to attract doctoral candidates. The first of these students was Frank Hogg, and (with or without status) I was to direct his research. Lectures began, informally at first, then more organized, and of course I had to lecture. The new Department called for a Chairman, a Professor. I could have done it; who knew the ropes better? But it was 'impossible'; the University would never permit it. Only a few years earlier, Theodore Lyman had refused to accept a woman as candidate for the PhD, and Shapley had somehow circumvented the difficulty. But this time it was not to be. I do not know what he tried to do, but he reported to me that President Lowell had said that 'Miss Payne should never have a position in the

University while he was alive'. Perhaps Shapley did make an attempt. But my nameless status remained nameless. Harry Plaskett was brought from Victoria to head the new Department.

As I look back, I see that this was a turning point in my career at Harvard. Plaskett had not expected to like me. Had I presumed to argue too vehemently with his father, the great J. S. Plaskett? Shapley had already paved the way for a difficult relationship by asking me 'how much it would disturb me' if Harry Plaskett were to come to Harvard Observatory? What did he say, I wonder, when he 'mentioned me to him'?

I remember the day when Plaskett arrived. Prompted by who knows what impulse, I had dressed with extra attention and put a blue ribbon in my hair. I remember his greeting: 'You're not at all like what I expected'. What *had* he expected? I wondered.

As it turned out, my relationship with him was not to be the same as with the visiting astronomers with whom I used to argue through the night. We became warm friends, but we never discussed astronomy. Scientifically we were on different wavelengths. He treated me as a woman, not as a scientist. I was not jealous of him, although the students assigned to me soon transferred their allegiance to him. I was sorry, but I considered that it was their loss; and it left me more time for research. Meanwhile I spent many happy hours in the Plaskett home, playing with their children and chatting with Mrs Plaskett. It was the first family I had got to know since I was grown up; I had very few human contacts in those days.

Only some years later, when Harry Plaskett was called to Oxford to succeed H. H. Turner, did I feel jealous of him. Of course I had no right to aspire to the Savilian Professorship, but I felt that I should have been as well qualified as he. Not for the first time, I felt I had been passed over because I was a woman.

When Plaskett left Harvard there was a search for a successor. Shapley said to me at this time: 'What this Observatory needs is a spectroscopist'. I replied indignantly that *I* was a spectroscopist, though I was being pushed against my will into photometry. I protested to no avail: a spectroscopist must be imported. The position was offered to Otto Struve, and he told me many years later why he had refused it. Shapley told him, he said, that 'Miss Payne shall give

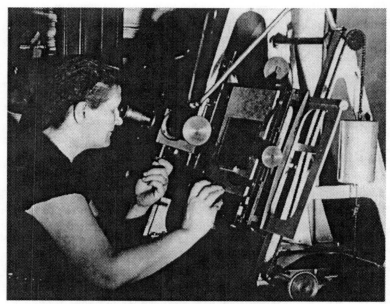
Fig. 14. Cecilia Payne-Gaposchkin at blink microscope, c. 1946

up spectroscopy', thus assuring him a free hand. He refused to accept the position on those terms. He had a noble, generous heart; he was one of the giants of his time. If only it had been my lot to work with him!

It was then that Donald Menzel was called to Harvard, after having made a name for himself at Lick Observatory. Again I was asked 'how much it would disturb me'? The groundwork for the 'Divide and Rule' system had been laid long before. It was not for many years, on Shapley's retirement, that I found that Menzel and I could form an alliance, rather than existing in a state of armed truce. This was a grave loss to me, and perhaps to science too. But now the situation was different from what it had been when Harry Plaskett was Professor: now I was associated with a man who treated me as a fellow-scientist.

Years passed and Lowell was no longer President of the University. Under James Conant the status of women at the Observatory underwent a change. Miss Cannon was as famous as any astronomer in the world, and justly so. For many years she had enjoyed the ambiguous title 'Curator of the Astronomical Photographs', which carried no status in the University. Now she was appointed Astronomer, and I received the same title. It was a step forward for me, for now I had a position, though still at a regrettable salary. My duties, research, lecturing, guidance of students, were actually those of Professor, but at least I now had a University position. I have spoken elsewhere of my admiration and affection for Professor King, who had been Phillips Professor (though actually he did little or no teaching). After his death I asked whether my title might be made 'Phillips Astronomer', and this was done.

Another lapse of years, another President of the University, and the time came for Shapley to retire as Director of the Observatory. After an agonizing time of indecision, Donald Menzel finally succeeded him. To Donald I owe the advancement that was finally accorded me. The finances of the Observatory had been a closely guarded secret, and when he learned what salary I had been getting, he told me that he was shocked. He promptly raised it, and soon doubled it. Moreover, he succeeded where Shapley had failed (though I shall never know how hard he had actually tried): I was made Phillips Professor and Chairman of the Department of Astronomy. Such was the generous treatment I was accorded by the man from whom I had been systematically estranged for many years. He did not let my sex, or my less-than-cooperative attitude, stand in my way.

The new position was no sinecure. I inherited a heavy teaching load and the responsibility for a large graduate school. There was literally no time for research, a setback from which I have never fully recovered. But as a mother sees her life renewed in her children, I saw my scientific efforts perpetuated in my students. There is no greater reward for an instructor than that of seeing students, whom we began by teaching, grow into fellow-scientists who end by teaching us. A few of the young men and women who were my responsibility during those hard years have given me this reward, and I thank them for it.

Since then my scientific life has been a crescendo of activity. More years have passed, and I have retired from Harvard University. Thence I moved to the affiliated Smithsonian Astrophysical Observatory, and as more years have passed I retired from that position also.

Fig. 15. Cecilia Payne-Gaposchkin c. 1950

But from Harvard Observatory I have never retired, and I hope that I shall never need to. There is more research to be done than there ever was; we know so much less than we did when I came there as a student more than 50 years ago.

As I look back I ask myself what difference it has made to me as a scientist that I was born a woman. As concerns the intellectual side of the matter, I should say that it has made very little. I am not conscious of having used any feminine wiles in connection with my scientific work; in fact I do not see how I could have done so. In that sense scientific work is inhuman. I will not accept the conclusions of another astronomer simply because I am fond of him, or reject them because I dislike him (though I admit there is a temptation here). Neither do I expect others to accept my arguments or praise my work because they like me, or to attack me because they do not. Some charming people do slipshod work, and some disagreeable people have made great contributions. I have said elsewhere that I wish that all scientific work might be anonymous.

On the material side, being a woman has been a great disadvantage. It is a tale of low salary, lack of status, slow advancement. But I have reached a height that I should never, in my wildest dreams, have predicted 50 years ago. It has been a case of survival, not of the fittest, but of the most doggedly persistent. I was not consciously aiming at the point I finally reached. I simply went on plodding, rewarded by the beauty of the scenery, towards an unexpected goal.

Young people, especially young women, often ask me for advice. Here it is, *valeat quantum*. Do not undertake a scientific career in quest of fame or money. There are easier and better ways to reach them. Undertake it only if nothing else will satisfy you; for nothing else is probably what you will receive. Your reward will be the widening of the horizon as you climb. And if you achieve that reward you will ask no other.

23

Science and myth

Dr Shapley handed me an unexpected experience when he suggested that I review a controversial and startling article. The subject has been so hotly debated that I must be exact. It was not a review of *Worlds in collision* that I was asked to write, but an appraisal of a pre-publication account of that widely read work. He had been asked to write it himself, and (very sensibly) tossed a hot potato into my lap.

I wrote an adverse review, which landed me in a lengthy and often comic controversy. Among other things, I was accused of having condemned a book unread; of course I had not read it, for it had not yet appeared. For that reason I later published a long review, based on careful reading of the work in question, in *Popular Astronomy*. I may say that I still stand by its contents.

The interrelationship between science and mythology has always interested me, and I think I may still have something to contribute to the subject. Certainly I do not approach the matter with a closed mind. *Worlds in collision* is a very well-written book that makes most fascinating reading. The only trouble with it is that it is wrong. My urbane adverse review would probably not have done much harm, but some other astronomers did not confine themselves to dispassionate criticism. Their angry reaction (in which I had no part) simply had the same effect as having a book 'banned in Boston'. It was certainly a factor in making the work a best seller. I am continually being asked to qualify and reassess my first judgement, but have always refused on the ground that, as we used to say in England, 'I don't want to send good time after bad'. I have never seen reason to modify my first opinion.

The problem is a much broader one than the question of whether

Science and myth

a particular book is 'right' or 'wrong', or of whether a 'wrong' book should be published. After looking the other way for many years I was at last compelled to come to grips with this problem. It had first come my way in the shape of spiritualism, and I had resolutely turned my face in the other direction. Dr Shapley had had his brush with spiritualism, and we have his word for it that all the spiritualists he had examined seemed to him to be frauds. But he did not succeed in proving this to the satisfaction of the public. I am glad that I was not involved in those investigations. I find in myself a persistent streak of extrasensory perception, and my Husband tells me that I am the most 'psychic' person he has ever known. Perhaps I am just unusually suggestible – an invaluable trait in an investigator, but one that must be tempered by an inflexibly critical spirit. I only know that psychical research is not for me.

If I were a member of the medical profession, I suppose I should be hostile to Christian Science. Yet how can I deny that faith healing is a fact? Who can fail to be impressed by a thinker and scientist of the stature of Alexis Carrel?

A physical scientist may not know enough to take a reasoned and informed position about spiritualism and faith healing. But experience with his own subject should surely make him reluctant to decide on the limits of possibility in these domains, so much more complex and difficult than his own. I cannot forget that I left biology for astronomy, because the latter seemed to be a more tractable subject.

Every scientist who is attempting to deepen and extend our understanding of the Universe knows that the keynote of science is progress. The ideas of yesterday are modified and replaced as new facts are discovered and woven into the pattern. When a province of thought abandons progress for dogma, it has come to a parting of the ways. When spiritualism and faith healing degenerate into cults they are no longer viable. This is the danger that underlies our attitudes to our own mysterious minds and bodies, and it tempts 'believers' into blind follies.

Other schools of thought come into more direct collision with astronomy. We can scarcely plead ignorance when we are confronted by the astrologer, the Flat Earth believer, and the votary of the UFO and the Chariots of the Gods.

In my early days, the Flat Earth group was much in evidence. It was interesting to learn that they had a pat answer to every astronomical argument, though it never seemed to me worthwhile to attempt a confutation. One might have expected that they would be silenced by the photography of the earth from satellites, but it is always possible to cry 'Fraud!'. And there is the unanswerable argument advanced by Edmund Gosse's father, so beautifully described in that writer's *Father and son*. It amounts to the contention that, in Alexander Wood's words, 'the Universe is a colossal practical joke on the part of the Creator'. We can only reply, in Eddington's words, that we do not believe that we live in that kind of universe. In the end, the question is one of faith.

There remains astrology, whose influence seems to be growing in an increasingly troubled world. Almost every newspaper carries a daily astrological page, and the Zodiac leaves its mark on all phases of life. Not long since, a casual acquaintance in the train asked me what work I did. After a little conversation she concluded: 'Ah, you are a *writer*'. 'And what', I enquired, 'are you?' She drew herself up proudly. 'I', she replied, 'am a *Libra*.'

I heard a discussion between a professional astrologer and an enquiring student. To the latter's frank question, how could she believe such nonsense, she replied serenely: 'What was good enough for Aristotle is good enough for me' (here, of course, she did grave injustice to Aristotle). It was a blind, if incorrect, appeal to dogma. Aristotle had a gigantic mind. If he were alive today we can be sure that he would be in the forefront of science, especially because he never lost sight of the importance of *facts*. But every scholar knows countless things that Aristotle did not know. Every housewife operates equipment of which he never could have dreamed. We scientists can indulge in sparkling fantasy when we imagine ourselves showing him our observatories and laboratories, and can picture his reactions and his delight.

Of course the stars have an influence on human life. It was astronomy, said Poincaré, that first taught mankind the existence of natural law. Astronomy led the way in navigation, in surveying the Earth, in the technology of projectiles (if this last, indeed, is to be reckoned a benefit). Astronomy has broadened the human horizon

Science and myth

from a parochial circle until it spans a Universe of variety and splendor that beggars the imagination, a Cosmos that continues to expand with the burgeoning resources of the human spirit.

But this is not what the astrologer wants to hear. His beliefs embrace an elaborate arithmetical system that links our view of the heavens to individual human destiny. He is very proud of his extensive calculations, as though a mere display of arithmetic were a guarantee of truth. But consider the Zodiac itself, those 12 'signs' that he associates with human fate. Our constellations are an accident of our position in space: move us 1000 light years and we should see a different pattern in the sky. The stars may look like a design traced on a surrounding sphere; we have only to look on them in three dimensions to realize that the legendary constellations are no more than an illusion.

The most obvious astronomical phenomenon, the seasons, has no doubt an effect on humanity. It is not difficult to believe that the physical conditions during the months of gestation have an effect on the unborn child. But it is a far cry from this to planetary and stellar influences. Nor is it unlikely that the phases of the moon, with their effects on the tides and the water table, may affect germination and the growth of crops.

Perhaps this sounds as though an astronomer has a closed mind in relation to astrology. That is not so. He is willing to accept conclusions based on adequate data and critical reasoning. He would respond to a massive statistical study, carried out with all the safeguards inherent in the proper treatment of observational data. But it hardly seems worthwhile to undertake such a study to test a group of ideas that have no rational basis. For these are dogmas inherited from an ignorant and credulous antiquity, and those who hold them are so dedicated that no amount of statistical discussion would shake their faith.

For *faith* is the key word. I have found consolation in the scholarly study by Seznec that traces the changing faces of the pagan gods through early and medieval times. I see in astrology nothing more than a debased survival of pagan religion. When we view it in the light of the development of human thought, it takes its rightful place. It was a noble science once, and as such it bore precious fruit, for it was the forerunner of astronomy. Only when it became a cult, when it fossilized, did it lose its value for mankind. Today it is no more than an intellectual strait-jacket.

The cult of the UFO is another facet of the same human urge. It is not new. The ninth-century Agobard of Lyons wrote: 'We have seen and heard many who . . . believe and assert that there is a certain region . . . whence ships come in the clouds: the which bear away the fruits of the earth . . . to that same country . . .'. Centuries later, Gervase of Tilbury recorded that 'folk came out of Mass one morning and saw an anchor let down from such a cloud-ship and grappled to a tomb. They heard the cries of the embarrassed cloud-sailors in the fog, until one came down the rope, hand by hand, and released the anchor . . . He gave up the ghost, stifled by the breath of our gross air . . . His fellows above, judging him to be wrecked, cut the cable, left their anchor, and sailed away . . .' In this circumstantial account, the facilities of the celestial navigators are understandably in tune with the times. Today the subject is discussed with such heat that I have no desire to add any fuel to the flames. The understanding of unexplained phenomena is, and ought to be, the goal of every scientist. In my own province of astronomy there are plenty of observations whose interpretation is still obscure. The UFO is a legitimate subject of enquiry, subject to the same tests of evidence and consistency as the quasar, the pulsar, and the X-ray burster. It should not be permitted to degenerate into a cult.

In the battle of words over *Worlds in collision* (which I mention because I have met it at first hand), 'the Scientists' were accused of wishing to suppress new ideas because they wanted to protect their own 'dogmas'. I can but suppose that the accusers had never read any scientific literature. For we spend our lives in trying to overthrow obsolete ideas and to replace them with something that represents Nature better. There is no joy more intense than that of coming upon a fact that cannot be understood in terms of currently accepted ideas. No excitement is comparable to that of devising, or learning of, a new theory. Einstein remarks somewhere that the finest fate for a scientific theory is to pave the way for a completer one, in which it survives as a special case. But every new fact must come under merciless scrutiny, every step in reasoning under meticulous criticism.

Only those who have shared in this activity can understand the joy of it. Science is a living thing, not a dead dogma.

It follows that no idea should be suppressed. That I totally disagree with what you say, but will defend to the death your right to say it, must be our underlying principle. And it applies to ideas that look like nonsense. We must not forget that some of the best ideas seemed like nonsense at first. The truth will prevail in the end. Nonsense will fall of its own weight, by a sort of intellectual law of gravitation. If we bat it about, we shall only keep an error in the air a little longer. And a new truth will go into orbit.

Finally, let us admit that dogma has its uses. It is not, it can never be, the doorway to new understanding. But we are not all scientists. Perhaps we should not quarrel with something that increases human satisfaction. If an idea makes life more bearable, perhaps it has made a valid contribution to the happiness of mankind.

24

Worlds not realized

When my Father left the academic world of Oxford and returned to the soil that his ancestors had tilled, he brought an intellectual heritage. 'At heart', he said, 'I am a Greek.' So it was that the first tales I heard were drawn, not from the Brothers Grimm, but from the Odyssey and the Argonauts.

But an Oxford man of the nineteenth century had close ties with the Church of England. As a Fellow of University College, he and a group of friends had accompanied the Chapel services with a string quartet. For years he acted as Organist in the Church at High Wycombe. He saw to it that his children were brought up in the Church of his fathers. I was baptized with water from the River Jordan. And when I was old enough, I was taken regularly to Wendover church.

There was a fascinating legend about the Church. It had been planned as a part of the village, but every night the stones were carried away by witches and piled up in the place where the Church now stands, almost a mile from the village. The field where the building was originally to be erected is still known as Wychell.

As I look back on myself before the age of 12, I see a ridiculously literal child. I took everything *au pied de la lettre*. When I heard a friend of my Mother say that her brother had forbidden her, 'on pain of death', to have her new dress made up by the local dressmaker, I looked at him with horror. Was he actually intending to murder his sister? And one heard such very odd things in Church. 'As I was musing, the fire kindled', said the thirty-ninth Psalm; what could it mean? Fires do not light themselves, and I was terrified of fire. Could one light a fire by thinking? Twice I had set myself on fire as a small child, once by trying to use a match as a pencil to draw with, and once by bringing my light, fuzzy hair too near a candle. I can still feel my Father's strong hands as he crushed the fire out. 'And thou, child, shalt be called the Prophet of the Highest': was this addressed to *all* children, and what did it mean? Perhaps it meant *me*.

As I grew older, my puzzlement began to give way to scepticism. The hymns that we sang expressed some thoughts that I felt I did not share; was it right to sing them under the circumstances? I have already described my statistical experiment with prayer. What really shook me, I think, was the blundering Sunday School lesson in which the curate was telling the story of the Good Samaritan, and got it all wrong. This seemed quite unpardonable. I fear that at that age I should have agreed with Marlowe's hero–villain that 'there is no sin but ignorance'. I had not yet realized that 'a little learning is a dangerous thing'; indeed it has been a life-long struggle to remind myself that *all* our learning is only 'a little' in comparison to what there is to learn.

Until my late teens I do not think I experienced anything like genuine reverence, and it struck unexpectedly. With an art class I was taken to the British Museum to 'draw from the antique'. It was during the First World War, and the most precious statues had been replaced by copies. Set to make a pencil sketch of the Demeter of Cnidos, I soon abandoned the pencil to gaze at her in rapturous adoration. She still seems to me the most beautiful creature I have ever looked on. Her picture has hung over my desk ever since, and it is on her face that I draw for inspiration. I tried my prentice hand at a sonnet, thinking perhaps of Swedenborg: 'All Nature is masculine; the Earth is the Mother'

Demeter

In thy sad face the universe I see,
    Mother and Bride of Nature! Those clear eyes
Gaze in rapt wonder on Eternity,
    Looking on thee I see it and grow wise.
About thee is the glory of the earth,
    Upon thy lips are all its joy and pain.

> To every thing that moves dost thou give birth,
>   And dead receiv'st them to thy breast again.
> Thou gleamest down the ages, in thy white
>   And dazzling splendour. Sure, no mortal planned
> Thy awful majesty, thy lovely might;
>   But God alone upon thee laid his hand.
> I could forget the world and all her lore
> To gaze upon that face for evermore.

Demeter laid a lasting spell on me. When the danger of war was over, the masterpiece of Scopas was brought back to her niche. It is, as everyone knows, in two pieces: the head of the finest marble, fitted into a body of coarser texture. One day I came upon my idol, ready to take her place again, the exquisite head lying apart on a piece of sacking. Perhaps I am the only living soul that has placed a passionate kiss on the lips of Demeter. I felt the same touch of inspiration that came to me many years later when I stooped to drink of the Castalian Spring in the great cleft above Delphi. I knew what my Father meant when he said: 'I am a Greek'.

Time passed, and I tried to come to grips with understanding the world. During my time at Cambridge it became an intellectual, rather than an emotional struggle. Knowledge seemed to be a terrible thing, leading to unbearable heights of isolation. I remember writing a short story about a man who understood everything, and was unable to survive the revelation. I encountered solipsist philosophy, and found myself unable to refute it. I turned to Eddington in my perplexity, and asked him whether he thought there were any arguments against it. He considered for awhile, as was his habit, and then answered: 'I do not anticipate that the universe will prove to be of that kind'. So this was the answer of the greatest intellect I have ever had the privilege to meet. I was willing to accept it. The best argument that I have ever been able to adduce is that I am so often wrong that I cannot picture the Universe as the figment of my extremely fallible intelligence.

Knowledge and understanding still seemed to me to lay a terrible burden on their possessor. If Demeter was the goddess of my adolescence, Prometheus was the god of my early maturity. Another essay in the most exacting form of verse dates from my early twenties

Prometheus

With fearless and untroubled glance he scanned
    The icy void and vastness of the sky,
    He saw the splendour with a prophet's eye,
And knew himself the savior; his the hand
To kindle at the stars a beacon brand.
    Too well he knew he might not hope to die,
    Foresaw the sharp unceasing agony
Reserved for him that dares to understand.

He chose to suffer, to rejoice in pain
    That buys the vaunted freedom for the race,
    The thunder of the god he had defied
Confronting with a proudly lifted face,
To view the writhing body with disdain,
    The flesh upon the spirit crucified.

Serenity has succeeded to these unquiet years. I see with surprise that Demeter has the face of the Mater Dolorosa, and Prometheus is the forerunner of a crucified Savior.

These pages may seem out of place in the record of the life of a scientist. They are written in answer to the common misconception that because an astronomer knows about the heavenly bodies, he must also be an authority on heaven. Nothing could be more untrue. And there is another misconception about which I have a word to say before I close. It is that a scientific man is necessarily a materialist.

It is true that we base our work on observed facts. If nothing were observed, there would be nothing to understand. But the facts are not the reality: that is something that lies beneath the facts and gives them coherence. If science, as I know it, can be described in a few words, it might be called a search for the Unseen. What really gives texture to the world is gravitation, is electromagnetism, is the entity that lies behind wave mechanics. The physical world has no actual substance at all; in my simplistic view its primal quality, the basis of its grandeur and majesty, is consistency. Primitive man worshipped

Sun and Moon because they were dependable, predictable. All our observations require that we dig deeper into understanding to show that they, too, are consistent with the pattern. Nature has always had a trick of surprising us, and she will continue to surprise us. But she has never let us down yet. We can go forward with confidence

> Knowing that Nature never did betray
> The heart that loved her.

Research

O Universe, o Lover
I gave myself to thee
Not for gold
Not for glory
But for love.
Our children are immortal,
I am the Mother.
The offspring of our love
Will bear the image of a humble mother
And also of a proud imperious Father.
Like Danae
I saw him in a stream of glowing stars;
Like Alkmena
Long, long I lay in his terrible embrace.
Their sons go striding round the firmament;
My children gambol at their heels.

Bibliography of works by Cecilia Payne-Gaposchkin

Notes on CPG bibliography
1. Where there are multiple authors of a paper, no designation of principal author has been made.
2. References were collected from an examination of the *Astronomisches Jahresbericht*, *Astronomy and Astrophysics Abstracts*, and the *Reader's Guide*.
3. As much as possible, the bibliography is in chronological order, but where there are multiple papers in one year, or where the year is not known (for some papers during the period of World War II), the chronological order is not exact.

Bibliography

Proper motions of the stars in the neighborhood of M36 (NGC 1960). *Monthly Notices of the Royal Astronomical Society*, **83**: 334, 1923

On the spectra and temperatures of the B stars. *Nature*, **113**: 783, 1924

On the absorption lines of silicon in stellar atmospheres. *Harvard College Observatory Circular*, no. 252, 1924

On ionization in the atmospheres of the hottest stars. *Harvard College Observatory Circular*, no. 256, 1924

A synopsis of the ionization potentials of the elements. *Washington National Academy Proceedings*, **10**: 322, 1924

On the spectra of class O stars. *Harvard College Observatory Circular*, no. 263, 1924

Stellar atmospheres. Harvard Monograph no. 1, Cambridge, England: W. Heffer & Sons, 1925

Astrophysical data bearing on the relative abundance of the elements. *Washington National Academy Proceedings*, **11**: 192, 1925 (Harvard Reprint 20)

The Balmer absorption series in stars of class A. *Harvard College Observatory Circular*, no. 287, 1925

Distribution of intensity in bright hydrogen lines of Gamma Cassiopeiae. *Harvard College Observatory Bulletin*, no. 837, 1926

On the source of certain lines in the spectrum of Gamma Cygni. *Harvard College Observatory Bulletin*, no. 841, 1926

(with Harlow Shapley) On the distribution of intensity in stellar absorption lines. *Proceedings of the American Academy of Arts and Science*, **61**, no. 10, July 1926 (Harvard Reprint 28)

Stellar evolution. *Science Monthly*, **22**: 419, May 1926 (Harvard Reprint 29)

Some applications of the ionization formula. *Washington National Academy Proceedings*, **12**: 717, 1926 (Harvard Reprint 33)

On the spectra of stars of class cF8. *Washington National Academy Proceedings*, **12**: 722, 1926 (Harvard Reprint 34)

Ionized vanadium in the solar spectrum. *Harvard College Observatory Bulletin*, no. 841, 1926

(ed. with Harlow Shapley) *Radio talks from the Harvard Observatory; the universe of stars*, by Harlow Shapley, Willem J. Luyten and others, Cambridge, Mass. The Observatory, 1926

On the astrophysical behavior of certain iron lines. *Harvard College Observatory Bulletin*, no. 836, 1926

Wolf–Rayet spectrum with absorption lines – CPD $-64°$ 1629. *Harvard College Observatory Bulletin*, no. 834, 1926

On nitrogen in the spectra of the sun and stars – oxygen in stellar atmospheres. *Harvard College Observatory Bulletin*, no. 835, 1926

Star with H-alpha bright – BD $+12°$ 4268. *Harvard College Observatory Bulletin*, no. 836, 1926

Residual intensities for the spectrum of Gamma Cygni. *Popular Astronomy*, **34**: 627, 1926

On the application of Milne's ejection theory to novae. *Harvard College Observatory Bulletin*, no. 843, 1927

Color magnitudes of seventeen stars near NGC 6231. *Harvard College Observatory Bulletin*, no. 848: part 7, 1927

Ten stars of class O. *Harvard College Observatory Bulletin*, no. 846: part 10, 1927

Twelve stars of class B showing emission lines. *Harvard College Observatory Bulletin*, no. 846: part 8, 1927

Spectroscopic energy diagrams for five stars of class O. *Harvard College Observatory Bulletin*, no. 844, 1927

On the spectra of southern stars of class O; first note. *Harvard College Observatory Bulletin*, no. 842, 1927

Spectrum line photometry for two stars of class O. *Harvard College Observatory Bulletin*, no. 843, 1927

On the spectra of southern stars of class O; second note. *Harvard College Observatory Bulletin*, no. 843, 1927

On the spectra of southern stars of class O; third note. *Harvard College Observatory Bulletin*, no. 844, 1927

(with C. T. Chase) The spectrum of supergiant stars of class F8. *Harvard College Observatory Circular*, no. 300, 1927

(with F. S. Hogg) Photometry of the spectrum of Mira Ceti at the maximum of 1926–27. *Harvard College Observatory Circular*, no. 308, 1927

(with F. S. Hogg) On methods in stellar spectrophotometry. *Harvard College Observatory Circular*, no. 301, 1927

The measurement of the intensity of spectrum lines, *Harvard College Observatory Circular*, no. 302, 1927

(with F. S. Hogg) A spectrophotometric study of the brighter Pleiades. I. The line intensities. *Harvard College Observatory Circular*, no. 303, 1927

(with F. S. Hogg) A comparison of line intensities derived from slit and objective prism spectra. *Harvard College Observatory Circular*, no. 304, 1927

Photometric line intensities for five late type stars. *Harvard College Observatory Circular*, no. 305, 1927

Photometric line intensities for normal and supergiant stars. *Harvard College Observatory Circular*, no. 306, 1927

On the interpretation of photometric line intensities. *Harvard College Observatory Circular*, no. 307, 1927

On the spectra of very luminous stars. *Popular Astronomy*, **36**: 292, 1928

(with H. Shapley) Spectroscopic evidence of the fall of meteors into stars. *Harvard College Observatory Circular*, no. 317, 1928

On the distortion of the continuous background by wide absorption lines. *Washington National Academy Proceedings*, **14**: No. 4, 296, April 1928 (Harvard Reprint 46)

A suggested spectral classification of O stars and of nebulae. *Harvard College Observatory Bulletin*, no. 855, 1928

On the contours of stellar absorption lines, and the composition of stellar atmospheres. *Washington National Academy Proceedings*, **14**: No. 5, 399, May 1928

(with F. S. Hogg) On methods and application in spectrophotometry. *Washington National Academy Proceedings*, **14**: No. 1, 88, Jan. 1928

(with F. S. Hogg) On the pressures in the atmospheres of stars. *Harvard College Observatory Circular*, 334, 1928

V Antliae, elements. *Harvard College Observatory Bulletin*, no. 860, 1928

X Antliae, elements. *Harvard College Observatory Bulletin*, no. 861, 1928

SY Aquarii, elements. *Harvard College Observatory Bulletin*, no. 861, 1928

ZZ Aquilae, elements. *Harvard College Observatory Bulletin*, no. 861, 1928

BE, BM, BP Aquilae, elements. *Harvard College Observatory Bulletin*, no. 861, 1928

Y, Z, RR, RS Arae, elements. *Harvard College Observatory Bulletin*, no. 860, 1928

RV, RZ Arae, elements. *Harvard College Observatory Bulletin*, no. 860, 1928

TT Arae, elements. *Harvard College Observatory Bulletin*, no. 861, 1928

Epsilon Aurigae, ongoing investigation of its spectrum. *Harvard College Observatory Bulletin*, no. 855, 1928

UW Aurigae, elements. *Harvard College Observatory Bulletin*, no. 861, 1928

RR Capricorni, elements. *Harvard College Observatory Bulletin*, no. 861, 1928

SV Carinae, elements. *Harvard College Observatory Bulletin*, no. 860, 1928

AO Centauri, elements. *Harvard College Observatory Bulletin*, no. 860, 1928

RY Centauri, elements. *Harvard College Observatory Bulletin*, no. 860, 1928

U Chamaeleontis, elements. *Harvard College Observatory Bulletin*, no. 860, 1928

RR Coronae austr., elements. *Harvard College Observatory Bulletin*, no. 860, 1928

T, U Doradus, elements. *Harvard College Observatory Bulletin*, no. 860, 1928

U Fornacis, elements. *Harvard College Observatory Bulletin*, no. 860, 1928

W, Y Indi, elements. *Harvard College Observatory Bulletin*, no. 860, 1928

TU Librae, elements. *Harvard College Observatory Bulletin*, no. 860, 1928

RX Lupi, elements. *Harvard College Observatory Bulletin*, no. 860, 1928

X, W Microscopii, elements. *Harvard College Observatory Bulletin*, no. 860, 1928

SY Monocerotis, elements. *Harvard College Observatory Bulletin*, no. 860, 1928

RU Octantis, elements. *Harvard College Observatory Bulletin*, no. 860, 1928

V Octantis, elements. *Harvard College Observatory Bulletin*, no. 860, 1928

X Octantis, elements. *Harvard College Observatory Bulletin*, no. 860, 1928

Z Octantis, elements. *Harvard College Observatory Bulletin*, no. 860, 1928

RR Pavonis, elements. *Harvard College Observatory Bulletin*, no. 860, 1928

RY Pavonis, elements. *Harvard College Observatory Bulletin*, no. 860, 1928

W Piscium, elements. *Harvard College Observatory Bulletin*, no. 860, 1928

SV Puppis, elements. *Harvard College Observatory Bulletin*, no. 860, 1928

BM Sagittarii, elements. *Harvard College Observatory Bulletin*, no. 860, 1928

DJ Sagittarii, elements. *Harvard College Observatory Bulletin*, no. 860, 1928

DN, DP, DQ Sagittarii, elements. *Harvard College Observatory Bulletin,* no. 860, 1928
BL Scorpii, elements. *Harvard College Observatory Bulletin,* no. 860, 1928
TU Scorpii, elements. *Harvard College Observatory Bulletin,* no. 860, 1928
WX Scorpii, elements. *Harvard College Observatory Bulletin,* no. 860, 1928
XX Scorpii, elements. *Harvard College Observatory Bulletin,* no. 860, 1928
YY Scorpii, elements. *Harvard College Observatory Bulletin,* no. 860, 1928
RY, SV Telescopii, elements. *Harvard College Observatory Bulletin,* no. 860, 1928
RU Trianguli austr., elements. *Harvard College Observatory Bulletin,* no. 860, 1928
W, Z Trianguli austr., elements. *Harvard College Observatory Bulletin,* no. 860, 1928
RW Velorum, elements. *Harvard College Observatory Bulletin,* no. 860, 1928
VZ Velorum, elements. *Harvard College Observatory Bulletin,* no. 860, 1928
WW Velorum, elements. *Harvard College Observatory Bulletin,* no. 860, 1928
T Volantis, elements. *Harvard College Observatory Bulletin,* no. 860, 1928
AP Centauri, elements. *Harvard College Observatory Bulletin,* no. 861, 1928
UU, XZ, AN Centauri, elements. *Harvard College Observatory Bulletin,* no. 861, 1928
Y, Z Crucis, elements. *Harvard College Observatory Bulletin,* no. 861, 1928
RW Librae, elements. *Harvard College Observatory Bulletin,* no. 861, 1928
SV Librae, elements. *Harvard College Observatory Bulletin,* no. 861, 1928
U Muscae, elements. *Harvard College Observatory Bulletin,* no. 861, 1928
Y Muscae, lichtkurve. *Harvard College Observatory Bulletin,* no. 861, 1928
RU Normae, elements. *Harvard College Observatory Bulletin,* no. 861, 1928
RX Pavonis, elements. *Harvard College Observatory Bulletin,* no. 861, 1928
AL, AM Sagittarii, elements. *Harvard College Observatory Bulletin,* no. 861, 1928
Anon. Sagittarii 181733, elements. *Harvard College Observatory Bulletin,* no. 861, 1928
BO, BR Sagittarii, elements. *Harvard College Observatory Bulletin,* no. 861, 1928
DL Sagittarii, elements. *Harvard College Observatory Bulletin,* no. 861, 1928
FN Sagittarii, elements. *Harvard College Observatory Bulletin,* no. 861, 1928
RU Sagittarii, elements. *Harvard College Observatory Bulletin,* no. 861, 1928

UZ, VZ Sagittarii, elements. *Harvard College Observatory Bulletin*, no. 861, 1928

BK Scorpii, elements. *Harvard College Observatory Bulletin*, no. 861, 1928

SZ Scorpii, elements. *Harvard College Observatory Bulletin*, no. 861, 1928

TY Scorpii, elements. *Harvard College Observatory Bulletin*, no. 861, 1928

TZ Scorpii, elements. *Harvard College Observatory Bulletin*, no. 861, 1928

VY Scorpii, elements. *Harvard College Observatory Bulletin*, no. 861, 1928

WW Scorpii, elements. *Harvard College Observatory Bulletin*, no. 861, 1928

TZ Tauri, elements. *Harvard College Observatory Bulletin*, no. 861, 1928

RR Telescopii, elements. *Harvard College Observatory Bulletin*, no. 861, 1928

RS Telescopii, lichtkurve. *Harvard College Observatory Bulletin*, no. 861, 1928

SW Telescopii, elements. *Harvard College Observatory Bulletin*, no. 861, 1928

RS Triang. austr., elements. *Harvard College Observatory Bulletin*, no. 861, 1928

RT, RU Velorum, elements. *Harvard College Observatory Bulletin*, no. 861, 1928

Observations bearing on Milne's generalization of the Saha theory. *Harvard College Observatory Bulletin*, no. 867, 1929

(with E. T. R. Williams) Photometry of hydrogen and calcium lines in stellar spectra. MN **89**, 526–558, 1929 (Harvard Reprint 55)

On the determination of the maximum intensities of stellar absorption lines. *Harvard College Observatory Bulletin*, no. 867, 1929

(with B. P. Gerasimovič) Note on the temperatures of F stars. *Harvard College Observatory Bulletin*, no. 866, 1929

V Pyxis, lichtkurve. *Harvard College Observatory Bulletin*, no. 868, 1929

The stars of high luminosity. Harvard Observatory Monographs No. 3. New York: McGraw-Hill, 1930

Summary of a spectrophotometric survey of the stellar sequence. *Harvard College Observatory Bulletin*, no. 874, 1930

On the spectra of the Wolf–Rayet stars. *Harvard College Observatory Bulletin*, no. 874, 1930

Classification of the O stars. *Harvard College Observatory Bulletin*, no. 878, 1930

(with P. ten Bruggencate) Effects of band spectra on the magnitudes and colors of stars. *Harvard College Observatory Bulletin*, no. 876, 1930

(with L. Campbell) On the nature of the period changes of long period variables. *Harvard College Observatory Bulletin*, no. 875, 1930

On the relation of period to mean density for cepheid variables. *Harvard College Observatory Bulletin*, no. 876, 1930

(with H. Shapley) The c-characteristic and the brightness of cepheid variables. *Harvard College Observatory Bulletin*, no. 872, 1930 (with L. Campbell) W Centaurus, visual and photographic lightcurves. *Harvard College Observatory Bulletin*, no. 872, 1930

(with L. Campbell) T Grus. *Harvard College Observatory Bulletin*, no. 872, 1930

(with L. Campbell) R Horologium. *Harvard College Observatory Bulletin*, no. 872, 1930

(with L. Campbell) R Octans. *Harvard College Observatory Bulletin*, no. 872, 1930

(with L. Campbell) S Octans. *Harvard College Observatory Bulletin*, no. 872, 1930

(with L. Campbell) S Pavo. *Harvard College Observatory Bulletin*, no. 872, 1930

(with L. Campbell) RS Scorpii. *Harvard College Observatory Bulletin*, no. 872, 1930

Photovisual magnitudes in Harvard Standard Regions. *Harvard College Observatory Bulletin*, no. 881, 1931

Photographic magnitudes of bright stars north of $+80°$. *Harvard College Observatory Bulletin*, no. 881, 1931

Notes on the Scorpio–Centaurus Cluster, I. The spectroscopic parallaxes. *Harvard College Observatory Circular*, no. 363, 1932

(with G. Maulbetsch) Notes on the Scorpio–Centaurus Cluster, II. The bright line stars and the problem of stellar rotation. *Harvard College Observatory Circular*, no. 364, 1931

Comparison of the Potsdam Photometry, and the Bonn and Cordoba Catalogues, with the Harvard Photovisual Photometry. *Harvard Annals*, **89** (3): 63-77, 1932

Photovisual magnitudes in Harvard Standard Regions. *Harvard Annals*, **89** (1): 1-39, 1931

An interpretation of absorption and emission lines in early type spectra. MN **92**, 368-88, 1932

(with C. J. Anger, G. Maulbetsch & G. W. Wheelwright) Notes on the Scorpio–Centaurus Cluster, III. Line intensities and the spectral classification of B stars. *Harvard College Observatory Circular*, no. 365, 1932

A heterochromatic study of the Pleiades. *Harvard Annals*, **89** (2): 43-59, 1932

Edward Skinner King. *Popular Astronomy*, XL: 2, Feb. 1932 (Harvard Reprint 76)

(with B. Gerasimovič) On the colors of SS Cygni variables. *Harvard College Observatory Bulletin*, no. 889, 3-6, 1932

Elements of the eclipsing variable U Gruis. *Harvard College Observatory Bulletin*, no. 889, 1932

The Harvard Photographic Photometry. *Harvard College Observatory Bulletin*, no. 892, 2-7, 1933

Relations between photovisual systems. *Harvard College Observatory Bulletin*, no. 892, 7-9, 1933

(with J. C. Boyce & D. H. Menzel) Forbidden lines in astrophysical sources, *Washington National Academy Proceedings*, **19**, 581-91, 1933

Absorption lines of Nv in stellar spectra. *Astrophysical Journal*, **77**, 299, 1933

A physical analysis of Wolf-Rayet spectra. *Washington National Academy Proceedings*, **19** (5): 492-4, May 1933 (Harvard Reprint 91)

(with Harlow Shapley) Photographic magnitudes of bright stars north of +57.5°. *Harvard Observatory Mimeograms*, Series I, No. 1, 1933

The analysis of the spectra of the Wolf-Rayet stars. *Zeitschrift für Astrophysik*, **7**, 1-21, 1933 (Harvard Reprint 96)

On the observation of SS Cygni variables. An appeal to variable star observers. *Astronomische Nachrichten*, **250**, 45-6, 1933

(with D. H. Menzel) On the interpretation of nova spectra. *Washington National Academy Proceedings*, **19** (7): 641-8, July 1933 (Harvard Reprint 95)

(with J. C. Boyce & D. H. Menzel) Additional identification of forbidden nebular lines. *Publications of the American Astronomical Society*, **7**, 214, 1933

(with F. W. Wright) Fifty-four photographic sequences in the Taurus region. *Harvard College Observatory Annals*, **89** (4): 81-90, 1934

Note on the spectra of semiregular variables. *Harvard College Observatory Bulletin*, no. 894, 16-17, 1934

(with D. H. Menzel) Nova Aquilae 3 (1918). *Publications of the American Astronomical Society*, **8**, 18, 1934

The Harvard Photographic Photometry +15° to +25°. *Publications of the American Astronomical Society*, **8**, 46, 1934 (first paper as C. Payne-Gaposchkin)

Red standards in Harvard Standard Regions. *Publications of the American Astronomical Society*, **8**, 46-7, 1934

(with J. C. Boyce & D. H. Menzel) Further identifications of nebular lines. *Publications of the Astronomical Society of the Pacific*, **46**, 213-215, 1934

They stand out from the crowd. *Literary Digest*, **118**: 11, 17 Nov. 1934

(with S. Gaposchkin & F. L. Whipple) Photovisual magnitudes of comparison stars for Nova Herculis. *Harvard College Observatory Bulletin*, 899, 1-5, 1935

Magnitudes of bright stars near Nova Herculis. *Harvard College Observatory Bulletin*, 898, 25-7, 1935

(with D. H. Menzel & F. L. Whipple) Spectrum. *Publications of the American Astronomical Society*, **8**, 112, 1935

(with S. Gaposchkin) A study of red color indices. *Harvard College Observatory Annals*, **80** (7): 1935 (with H. Shapley) Photographic magnitudes of bright stars between $+25°$ and $+15°$. *Harvard Observatory Mimeograms*, Series I, No. 2, 1935

(with S. Gaposchkin) Photographic magnitudes in selected areas at $-15°$. *Harvard Observatory Mimeograms*, Series II, No. 1, 1935

(with S. Gaposchkin) Photographic magnitudes of stars in selected areas at $-45°$. *Harvard Observatory Mimeograms*, Series II, No. 2, 1935

(with H. Shapley) A catalogue of photographic magnitudes of bright stars between declinations $+25°$ and $+15°$. *Harvard College Observatory Bulletin*, no. 899, 28-9, 1935

(with S. Gaposchkin) Photographic magnitudes of stars in selected areas at declination $-15°$. *Harvard College Observatory Bulletin*, no. 900, 22, 1935

(with S. Gaposchkin) Photographic magnitudes of eighteen thousand stars in selected areas. *Harvard College Observatory Bulletin*, no. 901, 25, 1935

(with S. Gaposchkin) Comparison of Harvard magnitudes in selected areas with the Cape Zone Catalogue. *Harvard College Observatory Bulletin*, no. 901, 25-30, 1935

The eclipsing star GO Cygni, B.D. $+34°$ 4095. *Harvard College Observatory Bulletin*, no. 898, 3-6, 1935

(with S. Gaposchkin) On the establishment of red standards by photographic methods. *Harvard College Observatory Annals*, **89** (5): 93-101, 1935

Color indices of giants and dwarfs. *Harvard College Observatory Annals*, **89** (6): 105-15, 1935

(with S. Gaposchkin) The scale of the Harvard-Groningen Durchmusterung in declination $-15°$. *Publications of the American Astronomical Society*, **8**, 112, 1935

Red standards at the North Pole. *Publications of the American Astronomical Society*, **8**, 142, 1935

(with H. N. Russell & D. H. Menzel) The classification of stellar spectra. *Astrophysical Journal*, **81**, 107-18, 1935

(with S. Gaposchkin) Photographic magnitudes of stars in selected areas at $-60°$. *Harvard Observatory Mimeograms*, Series II, No. 3, 1936

(with S. Gaposchkin) A study of red color indices. *Harvard College Observatory Annals*, **89** (7): 119-22, 1936

(with S. Gaposchkin) Comparison of the Harvard photographic magnitudes between $+20°$ and $+25°$ with the magnitudes of the Yale Zone Catalogue. *Harvard College Observatory Bulletin*, no. 902, 3-5, 1936

(with S. Gaposchkin) Comparison of the Cape Zone Catalogue and the Yale Zone Catalogue ($+25°$ to $+20°$) with Harvard Photographic Catalogues. *Publications of the American Astronomical Society*, **8**, 210, 1936

Note on the spectrum of Z Centauri. *Astrophysical Journal*, **83**, 173-6, 1936 (Harvard Reprint 123)

On the spectrum of the supernova S Andromedae. *Astrophysical Journal*, 83, 245–51, 1936

Note on the spectroscopic characteristics of the R Coronae Borealis type stars. *Harvard College Observatory Bulletin*, no. 903, 35–6, 1936

On the physical condition of the supernovae. *Washington National Academy Proceedings*, 22, 332–6, 1936 (Harvard Reprint 125)

(with F. L. Whipple) Applications of the theory of continuous ejection to Nova Herculis. *Harvard College Observatory Circular*, no. 413, 1936

(with F. L. Whipple) The early stages of Nova Herculis, I. The continuous and absorption spectra. *Harvard College Observatory Circular*, no. 412, 1936

(with F. L. Whipple) The early stages of Nova Herculis. *Publications of the American Astronomical Society*, 8, 227, 1936

(with F. L. Whipple) On the bright line spectrum of Nova Herculis. *Washington National Academy Proceedings*, 22, 195–200, 1936 (Harvard Reprint 121)

A reduction of the Göttingen Aktinometrie to the International Photographic System. *Harvard Observatory Mimeograms*, Series I, No. 3, 1937

Red magnitudes in Standard Regions at $+15°$. *Harvard College Observatory Annals*, 89, 123–42, 1937

New stars. *The Telescope*, 4, 100–6, 1937

(with F. L. Whipple) Line profiles in the spectrum of Nova Herculis. *Observatory*, 60, 46–47, 1937

(with F. L. Whipple) The early stages of Nova Herculis, II. The bright line spectra. *Harvard College Observatory Circular*, no. 414, 1937

Visual magnitudes of bright stars reduced to the Harvard Photovisual System 12^h to 23^h. *Harvard Observatory Mimeograms*, Series III, No. 2, 1938

Visual magnitudes of bright stars reduced to the Harvard Photovisual System 0^h to 11^h. *Harvard Observatory Mimeograms*, Series III, No. 1, 1938

On the reduction of certain visual catalogues to the Harvard Photovisual System. *Harvard College Observatory Annals*, 89 (12): 191–9, 1938

Red indices of stars in eight southern selected areas. *Tercentenary Papers, Harvard College Observatory Annals*, 105, 383–95, 1938

New catalogues of photographic and visual magnitudes. *Harvard College Observatory Bulletin*, no. 907, 39, 1938

(with D. H. Menzel) Investigations of the spectra of novae, I. Intensities of bright lines in the spectrum of Nova Pictoris. *Harvard College Observatory Circular*, no. 428, 1938

(with S. Gaposchkin) *Variable stars*. Harvard Observatory Monographs, no. 5, Cambridge, Mass, 1938

Red indices in southern selected areas. *Astrophysical Journal*, 90, 321–51, 1939 (Harvard Reprint 181)

Variable stars: a plan of study. *Proceedings of the American Philosophical Society*, **81**, 189–210, 1939 (Harvard Reprint 170)

(with F. L. Whipple) Spectrophotometry of Nova Herculis in its intermediate stages. *Harvard College Observatory Circular*, no. 433, 1939

(with F. L. Whipple) Synthetic spectra for supernovae. *Washington National Academy Proceedings*, **26**, 264–72, 1940 (Harvard Reprint 196)

Miss Cannon and stellar spectroscopy. *Telescope*, **8**, 62–3, 1941

The topography of the Universe. *Telescope*, **8**, 112–14, 1941

Annie Jump Cannon. *Science*, **93**, 443–4, 1941

A new angle on an old friend (on HD 193576, a spectroscopic double star). *Telescope*, **8**, 68–70, 1941

(with S. Gaposchkin) Interesting variable stars. *Popular Astronomy*, **49**, 311–19, 1941

A photographic study of RU Camelopardalis. *Harvard College Observatory Bulletin*, no. 915, 10–14, 1941

The system of VW Cephei. *Publications of the American Astronomical Society*, **10**, 127, 1941

Etude spectrophotométriques des novae. In *Les novae et les naines blanches*, ed. A. J. Schaler, pp. 69–91, Colloque international d'astrophysique tenu au Collège de France 17–23 juillet 1939, 1941

(with F. L. Whipple) Theoretical synthesis of supernovae spectra. *Proceedings of the American Philosophical Society*, **84**, 1–32, 1941 (Harvard Reprint 223)

(with S. Gaposchkin) A spectrophotometric study of the bright novae. *Harvard College Observatory Circular*, no. 445, 1942

(with S. Gaposchkin) On the material ejected from novae. *Washington National Academy Proceedings*, **28**, 482–90, 1942 (Harvard Reprint 246)

Novae and related stars, at the Inter-American Astrophysical Conference, 16–25 Feb. 1942, *Popular Astronomy*, **50**, 140–50, 1942

(with S. Gaposchkin) On the dimensions and constitution of variable stars. *Washington National Academy Proceedings*, **28**, 490–5, 1942 (Harvard Reprint 247)

(with V. K. Brenton) A study of the RV Tauri variables. *Washington National Academy Proceedings*, **28**, 496–500, 1942 (Harvard Reprint 248)

(with V. K. Brenton & S. Gaposchkin) The variables of RV Tauri type, *Harvard College Observatory Annals*, **113**, 1–65, 1943

(with S. Gaposchkin) Variable stars and the sources of stellar energy. *Proceedings of the American Philosophical Society*, **86**: 3, 1943 (Harvard Reprint (2): 2)

Variable stars in Milton field 54. *Harvard College Observatory Annals*, **115**, 1–10, 1943–46

Variable stars in Milton field 52. *Harvard College Observatory Annals*, **115**, 25–46, 1943–46

Variable stars in Milton field 49. *Harvard College Observatory Annals*, **115**, 87-102, 1943-46

Variable stars in Milton field 47. *Harvard College Observatory Annals*, **115**, 113-16, 1943-46

Variable stars in Milton field 45. *Harvard College Observatory Annals*, **115**, 121-7, 1943-46

Problems of stellar evolution. *Sky and Telescope*, **2** (9): 5-7, 1943

On magnitudes and colors of stars north of $+80°$. *Astrophysical Journal*, **97**, 78-9, 1943(?)

The eclipsing star AR Monocerotis. *Astrophysical Journal*, **100**, 251-4, 1944

(with S. Gaposchkin) Eclipses of stars with thick atmospheres. *Astrophysical Journal*, **101**, 56-70, 1945

(with S. Gaposchkin) Relation between components of binary stars. *Astronomical Journal*, **52**, 29-31, 1946

(with C. Boyd) The light curves of R Aquarii. *Astrophysical Journal*, **104**, 357-61, 1946

The light-curves of Z Andromedae and AX Persei. *Astrophysical Journal*, **104**, 362-9, 1946

The system of RX Cassiopeiae. *Astrophysical Journal*, **103**, 299-309, 1946

(with F. W. Wright) The photographic light-curve of T Coronae Borealis. *Astrophysical Journal*, **104**, 75-81, 1946

J. Jeans, *The Universe around us*. *Sky and Telescope*, 4 (2): 14-15, 1943-46

On J. Jeans. *Sky and Telescope*, **6** (1): 10, 1946

On photometric atlas of stellar spectra of Hiltner. *Sky and Telescope*, **6** (1): 15, 1946

Harlow Shapley (with a biographical sketch of the author). *Science Monthly*, **64** (3): 189, March 1947

(with S. Gaposchkin & M. Mayall) The variations of Rho Cassiopeiae. *Astronomical Journal*, **52**, 123, 1947

The remarkable variable Y Ceti. *Astronomical Journal*, **52**, 123, 1947

On the light curves of cepheid variables. *Astronomical Journal*, **52**, 218-26, 1947 (Harvard Reprint (2) 18)

The Perkins spectrograms of Gamma Cassiopeiae. *Astronomical Journal*, **53**, 198, 1948

The connection of motion with intrinsic variability. *Astronomical Journal*, **53**, 193-4, 1948

(with S. Gaposchkin) Variable stars, a study completed. *Science*, **107**, 590, June 4, 1948

Phase relations for intrinsic variables. *Astronomical Journal*, **54**, 185, 1949

(with F. L. Whipple) Synthetic spectra for supernovae. *Publications of the American Astronomical Society*, **9**, 267-8, 1949 (1939)

Old Sol's balance sheet! *Science Digest*, **25**, 61-6, 1949

Most important star. *Christian Science Monitor*, May 7, 1949

I. Velikovsky, *Worlds in collision. Popular Astronomy*, **58**, 278–86, 1950
G. Abetti, *Storia dell' astronomia. Sky and Telescope*, **9**, 142–3, 1950
C. F. von Weizsäcker, *The history of nature. Sky and Telescope*, **9**, 194, 1950
W. H. McCrea, *Physics of the Sun and stars. Sky and Telescope*, **10**, 123, 147, 1950
C. E. P. Brooks, *Climate through the ages. Sky and Telescope*, **9**, 302, 1950
Variable stars in Milton field 43. *Harvard College Observatory Annals*, **115**, 137–55, 1950
Variable stars in Milton field 41. *Harvard College Observatory Annals*, **115**, 173–80, 1950
Variable stars in Milton field 39. *Harvard College Observatory Annals*, **115**, 189–96, 1950
Variable stars in Milton field 37. *Harvard College Observatory Annals*, **115**, 205–12, 1950
Variable stars in Milton field 35. *Harvard College Observatory Annals*, **115**, 219–23, 1950
Variable stars in Milton field 33. *Harvard College Observatory Annals*, **115**, 229–33, 1950
Variable stars in Milton field 31. *Harvard College Observatory Annals*, **115**, 249–56, 1950
Variable stars in Milton field 29. *Harvard College Observatory Annals*, **115**, 265–71, 1950
E. Whittaker, *A history of the theories of aether and electricity, the classical theories. Sky and Telescope*, **11**, 63, 1951
The intrinsic variable stars. In *Astrophysics*, New York: McGraw-Hill, ed. J. A. Hynek, 1951
The nova phenomenon. In *Problems of cosmical aerodynamics*, J. M. Burgers & H. C. van de Hulst (eds.), Chapter 15, 107–15, Dayton, Ohio: Central Air Documents Office, 1951
(with G. C. McVittie) A model of a spiral galaxy. MN **111**, 506–22, 1951
Stars in the making. Cambridge: Harvard University Press, 1952
Worlds in collision. *Proceedings of the American Philosophical Society*, **96** (5): 519–25, 1952 (Harvard Reprint (2) 41)
E. J. Holmyard, *British Scientists. Sky and Telescope*, **11**, 93–4, 1952
O. Neugebauer, *The exact sciences in antiquity. Sky and Telescope*, **11**, 174, 227, 1952
E. A. Milne, *Sir James Jeans. Sky and Telescope*, **12**, 134 and **13**, 20, 1952
Variable stars and galactic structure. *Nature*, **170**, 223–5, 1952 (Harvard Reprint 364)
(with S. Gaposchkin) Variable stars in Milton field 27. *Harvard College Observatory Annals*, **118**, 1–3, 1952
(with S. Gaposchkin) Variable stars in Milton field 25. *Harvard College Observatory Annals*, **118**, 15–24, 1952

(with S. Gaposchkin) Variable stars in Milton field 23. *Harvard College Observatory Annals*, **118**, 33–9, 1952

(with S. Gaposchkin) Variable stars in Milton field 21. *Harvard College Observatory Annals*, **118**, 45–51, 1952

(with S. Gaposchkin) Variable stars in Milton field 19. *Harvard College Observatory Annals*, **118**, 71–9, 1952

(with S. Gaposchkin) Variable stars in Milton field 17. *Harvard College Observatory Annals*, **118**, 89–98, 1952

(with S. Gaposchkin) Variable stars in Milton field 15. *Harvard College Observatory Annals*, **118**, 103–4, 1952

(with S. Gaposchkin) Variable stars in Milton field 13. *Harvard College Observatory Annals*, **118**, 111–17, 1952

(with S. Gaposchkin) Variable stars in Milton field 11. *Harvard College Observatory Annals*, **118**, 131–40, 1952

(with S. Gaposchkin) Variable stars in Milton field 9. *Harvard College Observatory Annals*, **118**, 147–54, 1952

(with S. Gaposchkin) Variable stars in Milton field 7. *Harvard College Observatory Annals*, **118**, 165–9, 1952

(with S. Gaposchkin) Variable stars in Milton field 5. *Harvard College Observatory Annals*, **118**, 179–80, 1952

Kourganoff, Vladimir: Detailed structure of the Galaxy and stellar association (translation). *Astronomical News Letter* no. 64. Cambridge, Mass., 1952

(with S. Gaposchkin) Variable stars in Milton field 3. *Harvard College Observatory Annals*, **118**, 179–80, 1952

(with S. Gaposchkin) Variable stars in Milton field 1. *Harvard College Observatory Annals*, **118**, 217–24, 1952

Why do galaxies have a spiral form? *Scientific American*, **189**: 34, 1953

Stars in the making. London: Eyre and Spottiswoode, 1953

Variable stars and galactic structure. London: University of London, the Athlone Press, 1954

The cepheid variables and RR Lyrae stars. *Harvard College Observatory Annals*, **113**, 153–86, 1954

The red variable stars. *Harvard College Observatory Annals*, **113**, 191–208, 1954

Introduction to astronomy. New York: Prentice-Hall, Inc., 1954

Introduction to astronomy. (English edition) London: Eyre and Spottiswoode, 1956

Population II cepheids of the Galaxy. In *Vistas in astronomy*, Arthur Beer (ed.), 1142–9, New York: Pergamon Press, 1956

Die Entstehung und Entwicklung der Sterne (translation of *Stars in the making*). Moscow: Government Printing House, 1956

Spectrophotometric study of stellar rotation, an analysis of Beta Cassiopeiae. Publ. ASP **69**, 46–53, 1957 (Harvard Reprint 454)

The galactic novae. New York: Interscience Publ., 1957

Le stelle in formazione (translation of *Stars in the making*). Milano: Feltrinelli, 1958

Relationships between variable stars whose spectra show bright lines. In *Etoiles à raies d'émission*, P. Swings (ed.), 483–93, Cointe-Sclessin, Belgium: Institut d'Astrophysique, 1958 (Communications presented at the 8th Colloque International d'Astrophysique held at Liège, 8 to 10 July 1957)

Cepheid variables and the period–luminosity relation. *Journal of the Washington Academy of Science*, **49**, 333–50, 1959 (Harvard Reprint 536)

The absolute magnitudes of RR Lyrae stars. In *The scale of the Galaxy: a symposium.* Publications of the Astronomical Society of the Pacific **73**, 100–2, 1961 (Harvard Reprint 586)

Variable stars in the small Magellanic cloud. *Astronomical Journal*, **66**, 293, 1961

O. Struve, *The Universe. Physics Today*, **16** (2): 64, 66, 1962

Stars and stellar systems, Compendium of Astronomy and Astrophysics, Vol. 2, Astronomical Techniques. *Physics Today*, **16** (1): 63, 1962

New survey of the variable stars in the small Magellanic Cloud. *Astronomical Journal*, **67**, 279, 1962

Otto Struve as an astrophysicist. *Sky and Telescope*, **25**, 308–10, 1963

(ed.) W. Baade, *Evolution of stars and galaxies.* New York: Oxford University Press; Cambridge, Mass.: Harvard University Press, 1963

(with S. Gaposchkin) Photographic observations of CE Cassiopeiae. *Publications of the Astronomical Society of the Pacific*, **75**, 171–5, 1963

The 1960 minimum of R Coronae Borealis. *Astrophysical Journal*, **138**, 320–41, 1963

The galactic novae. New York: Dover Publ. Inc., 1964

The period–luminosity relation in the small Magellanic cloud. In *The position of variable stars in the Hertzsprung-Russell diagram, Kleine Veröffentlichungen der Remeis-Sternwarte Bamberg*, **4** (40): 178–81, 1965

Evolución del sistema solar. *Revista Astronomia*, **37** (162): 5–9, 1966

(with S. Gaposchkin) Variable stars in the small Magellanic Cloud. *Smithsonian Contributions to Astrophysics*, no. 9, 1966

(with S. Gaposchkin) Relation of light curve to period for cepheids in the small Magellanic Cloud. In *Vistas in astronomy*, vol. 8, eds. A. Beer & K. A. Strand, 191–201. Oxford: Pergamon Press, 1966

Orbital resonance as a possible source of nova outbursts. *Bulletin of the American Astronomical Society*, **1**, 189, 1969

Period and velocity curve of AE Aquarii. *Astrophysical Journal*, **158**, 429, 1969

(with K. Haramundanis) *Introduction to astronomy*, 2nd edn, Englewood Cliffs, N.J.: Prentice-Hall, Inc., 1970

The variable stars of the large Magellanic Cloud. *Smithsonian Contributions to Astrophysics*, no. 13, 1971

Comparison of the cepheid variables in the Magellanic Clouds and the Galaxy. In *The Magellanic clouds*, A. B. Muller (ed.), Astrophysics and Space Science Library, vol. 23, 34–46, Dordrecht, Holland: D. Reidel Publ. Co., 1971

Myth and Science. *Journal for the History of Science*, **iii**: 206–11, 1972

Cepheid variables in large and small Magellanic Clouds. In *Galactic astronomy*, vol. 1, Hong-Yee Chiu & A. Muriel (eds.), 245–329. New York: Gordon and Breach Science Publ., 1972

(with K. Haramundanis) Intrinsic ultraviolet colors from OAO-II Celescope observations. *Bulletin of the American Astronomical Society*, **4**, 331, 1972

(with K. Haramundanis) Intrinsic ultraviolet colors from OAO-II Celescope observations for stars on the main sequence. *Astronomical Journal*, **78**, 395–400, 1973

Time table for star formation in the large Magellanic Cloud. In *Stellar ages*, Proceedings of the IAU Colloquium, no. 17 (III): 1–12, 1973

F. Zwicky, a special kind of astronomer. *Sky and Telescope*, **47**, 311–13, 1974

(with E. Avrett, R. Davis, W. Deutschman, K. Haramundanis, R. Kurucz, E. Peytremann & R. Schild) Report on the Celescope ultraviolet observations from the OAO-2 satellite and associated research at the Smithsonian Astrophysical Observatory. In *Reports of new ultraviolet observations*, Astronomical Institute of Utrecht, Internal Report ROF72, pp. C2.1–19, 1974

Distribution and ages of Magellanic cepheids. *Smithsonian Contributions to Astrophysics*, no. 16, 1974

Period, color, and luminosity for cepheid variables. *Smithsonian Contributions to Astrophysics*, no. 17, 1974.

(with C. A. Whitney) Analysis of broad-band photometry of the long-period variables. *Smithsonian Astrophysical Observatory Special Report*, no. 370, 1976

(with C. A. Whitney) Analysis of broad-band photometry of the long-period variables. 20ᵉ Colloque international d'Astrophysique held at Liège, in *Astrophysique et spectroscopie, Mémoires de la Société royale des sciences de Liège, Collect. 8°, 6ᵉ Ser.*, **9**: 419–35, 1975

Comparison of colour curves of Mira stars of spectrum M and S. In *Abundance effects in classification*, IAU Symposium no. 72: 91–3. Dordrecht, Holland: D. Reidel Publ. Co., 1976

Past and future novae. In *Novae and related stars*, ed. M. Friedjung, pp. 3–32. Dordrecht, Holland: D. Reidel Publ. Co., 1977

Russell and the composition of stellar atmospheres. *Dudley Observatory Report*, no. 13: 15–18, 1977

Fifty years of novae. (Henry Norris Russell Prize Lecture of the American Astronomical Society), *Astronomical Journal*, **82**, 665–73, 1977

The development of our knowledge of variable stars. *Annual Review of Astronomy and Astrophysics*, **16**, 1–13, 1978

Stars and clusters. Cambridge, Mass.: Harvard University Press, 1979

The Garnett letters, privately printed, 1979

The dyer's hand, privately printed, 1979

Kontact mit den sternen, by Reinhard Breuer (translation) Breidenstein KG: Umschau Verlag; English publisher: Reading, England: W. H. Freeman and Co. Ltd., 1979

(with Susan G. Kleinmann) The reddest stars in the two micron sky survey. *Earth and extraterrestrial Sciences*, **3**, 161–71, 1979

Postlude

Cecilia Helena Payne was born on 10 May 1900 at Holywell Lodge in Wendover Buckinghamshire, England, the elder daughter of Edward John Payne of High Wycombe, England and Emma Leonora Helena (Pertz) Payne of Coblenz, Prussia. She attended the local school in Wendover, St. Mary's College, Paddington, 1916–18 and St. Paul's Girls' School, Brook Green, Hammersmith, 1918–19, and received the Mary Eward Scholarship for Natural Sciences to Newnham College, Cambridge University in 1919. At Newnham she studied botany, physics and chemistry, receiving her BA degree in 1923. In the same year she became a National Research Fellow at Harvard University, Cambridge, Massachusetts, USA, and became affiliated with the Harvard College Observatory.

In 1924 she was elected to membership of the American Astronomical Society and in 1925 she received her PhD from Radcliffe College, the first graduate student in astronomy to receive a degree from the Harvard College Observatory. Her doctoral thesis *Stellar atmospheres* won wide acclaim. In 1928 she became a member of the International Astronomical Union. She became a US citizen in 1931 and in 1934 she married Sergei I. Gaposchkin, of Eupatoria, Crimea, Russia. Their union produced three children: Edward (1935), Katherine (1937) and Peter (1940).

In 1934 she was the first recipient of the Annie J. Cannon Prize of the American Astronomical Society, a prize awarded for significant contributions to astronomy. She was elected to membership in the prestigious American Philosophical Society in 1936, and in 1938 she became Phillips Astronomer of the Harvard Observatory.

She received many honorary degrees and medals for her contri-

Fig. 16. Cecilia Payne-Gaposchkin receiving Rittenhouse Medal 1961 (Courtesy of the Franklin Institute Science Museum, Philadelphia, PA)

butions to science: honorary DScs from Wilson College (1942), Smith College (1943), Western College (1951), Award of Merit from Radcliffe College (1952), 28th Joseph Henry Lecture at the Philosophical Society, Washington, D.C. (1958), honorary DSc Colby College (1958), Rittenhouse Medal of the Franklin Institute (1961), honorary DSc from Women's Medical College of Philadelphia (1961). From 1945 to 1947 she was President of the Radcliffe Chapter of the Society of Sigma Xi, and in 1952 she received her MA and DSc from Cambridge University, England.

In 1956 she became the first woman to be advanced to the rank of professor at Harvard and the first woman department chairman at that institution. She held these positions until 1966.

From 1967 she was Emeritus Professor of Harvard University and became affiliated with the Smithsonian Astrophysical Observatory, later the Center for Astrophysics. In 1976 she received the Henry Norris Russell Prize of the American Astronomical Society. She was the first woman to receive the Prize, given for eminence in astronomical research. In 1977 she received an unusual accolade when the minor planet 1974 CA, discovered at the Harvard Observatory's Agassiz Station, was named Payne-Gaposchkin in her honor.

She contributed an article to *Annual Review of Astronomy and Astrophysics* in 1978. In this article, 'The development of our knowledge of variable stars', she summarizes how knowledge of variable stars accumulated during the twentieth century, and illustrates her discussion with examples from individual stars, her well-known friends. She demonstrates her great gift for the assimilation of vast amounts of material and the facility with which she could pull all the facts together and make reasonable sense out of a myriad details. She closes her discussion by directing the route of new research toward the Magellanic Clouds, 'the happy hunting ground for the student of variable stars'.

She was a member of many learned societies, among them the American Astronomical Society, the American Philosophical Society, the American Academy of Arts and Sciences, the International Astronomical Union, the Royal Astronomical Society, Phi Beta Kappa, and Sigma Xi. She wrote eight major books on astronomical subjects, among them *Stars in the making, Galactic novae, Introduction to Astronomy* and *Stars and clusters*. She died in her 80th year.

Index

In this index the letters n and p after page nos. refer to notes and photographs, respectively.

100-inch reflector 55
1974CA, minor planet 258
AAUW 32
Abelard and Heloise 63
Åbo (Turku, Finland) 192
absolute dimensions (of stars) 215
absolute luminosities (of stars) 215
absolute magnitude 22
absolute masses (of stars) 215
absorption lines viii
Abt, H. A. (b. 1925), German-born US astronomer, spectroscopist xi
abundance of elements in stellar atmospheres 19
abundance of hydrogen and helium, solar 20
abundance of metals 18
actress 43
Adams Prize Essay 14, 31, 165
Adams, Walter Sydney (1876–1956), US astronomer, Director of Mount Wilson Observatory; spectroscopist xi, 6, 24, 35, 183
Admiralty 220
Advanced Physics Laboratory 220
advisor, thesis viii
AE Aquarii (variable star) 84
Agassiz Station, Harvard, Mass. 258
Agobard of Lyons (769/779–840), Archbishop, writer on theological and liturgical subjects 232
Aitken, Robert G. (1864–1951), US astronomer, Director of Lick Observatory; worked on double stars 17, 32, 166
Alamagordo, New Mexico 54
Alaska, visit to 64
Albert Hall 105

Albert, Eugen d' (1864–1932), Scottish–born pianist and composer 82
Aldebaran (star) 147
All's well that ends well xixn
Alnitam (star) 147
America (*see also* United States) 175
American Academy of Arts and Sciences 258
American Association of University Women 17
American Association of Variable Star Observers 152
American Astronomical Society xii, xx, 17, 45, 64, 208, 256
American Institute of Physics xvii, 29
American Men of Science 23
American Philosophical Society 10, 27, 37, 45, 256
Ames, Adelaide (1900–1932), US astronomer; worked on galaxy counts 27, 29, 138p, 142, 154, 163, 189, 191, 208, 212
analysis of the lines of nitrogen 23
ancient Greek language 40
Anderson, John A., US astronomer 184
Andromeda Galaxy Cepheids ix, 209
Andromeda Nebula x
Andromeda Nebula, Cepheid variables in ix
Andromeda Spiral 119
Annie J. Cannon Prize, American Astronomical Society xi, 256
Annual Reviews of Astronomy and Astrophysics xvi, 258
anomalous excitation viii
Antigone 102

259

Index

Antony 156
appointments for women from Harvard 25
Arber, Agnes (1879–1960), British biologist; contributed to philosophy of biological study; wrote on botany and its history 114
Arc de Triomphe 204
Archimedes 178, 208
argonauts 234
Argus 90
Aristotle 3, 114, 230
Arp, Halton C. (b. 1927) US astronomer; worked on globular clusters, novae, evolution of galaxies, quasars 215
Aston, Francis William (1877–1945), English physicist; invented mass spectrograph 115
astrologer 229
Astronomical Society of the Pacific xvi
Astronomisches Gesellschaft 54, 195
Astronomy 20, graduate research course 29
Astrophysical Journal 183
Atlantic Ocean 129
atmosphere, Sun's 18
attractiveness xiv
Attwood-Mathews, Florence 81
Attwood-Mathews, St. John 81
Austin, Texas xi
Australia 64
Austria, travels in 52, 61
Award of Merit, Radcliffe College 257
Aylesbury, England 77

B stars 169
Baade, Walter (1893–1960), German astronomer; developed theory for Population I, II stars; worked on galaxies, cosmology 119, 203, 204p
Bach, Johann Sebastian 94, 108
Bailey, Solon (1854–1931), US astronomer; placed highest scientific station (El Misti, 19 000 ft); worked on meteors, photometry, variables 30, 138, 151, 157, 207
Baker, John Gilbert (1834–1920), British botanist; worked in horticulture, wrote monographs on ferns, roses 113

Bakst, Léon N. (1866–1924), Russian artist, set and costume designer 106
Balanowsky, Mrs 54
Baltic Sea 192
Barbier, Daniel 203
Barnard College, Columbia University 26, 27, 36
Bateson, William (1861–1926), English biologist; worked in genetics 113
Bath, England 56
Batory, SS 55
Baum, William A. (b. 1924), US astronomer; worked in photoelectric photometry, globular clusters, cosmology, development of optical instruments 151
Bayswater, London 96
Beals, Carlyle S. (b. 1899), Canadian astronomer; interpreted spectra of Wolf–Rayet stars, studied meteoritic and lunar craters 203, 204p
Beaudoin, Doris 50
Beethoven, Ludvig van 82
Belgium 52
Belopolsky, Aristarch A. (1854–1934), Russian astronomer; discovered several spectroscopic binaries, investigated variables, planetary rotation 53, 194
Benn, William Wedgwood (1877–1960), British political leader 83
Berlin, Germany 59, 79, 195
Beta Lyrae 140, 201
Bethe, Hans A. (b. 1906), German-born US theoretical physicist 182
Bible, The 10
Big Bang 3
binaries, eclipsing viii, ix
binary stars in clusters ix
binary, spectroscopic x
Blake, William 81, 193
blink microscope 215, 224p
Boddington Wood 86
Boethius, A. Manlius Severinus 114
Bohr theory of atom 2, 21
Bohr, Niels (1885–1962), Danish physicist; contributed to quantum and nuclear theory 7, 116
Bok family (family of Bart J. Bok) 191

Bok, Bart J. (1906–1992), Dutch-born US astrophysicist, discovered Bok's globules in nebulae 48
Boston, Massachusetts 61
botany, love of 63
Boulder Dam, Nevada 54
Boulogne, France 59
Bowie, Frank 143, 180
Boyd, L. G. 30
Boyden station, South Africa 25, 36
Boyden telescope 33
Bragg, William (1890–1971); British physicist; worked on crystal structure and X-rays 220
Brahe, Tycho, his observatory 53
Bredichin, Fyodor Aleksandrovich (1831–1904), Russian astronomer; worked on comets, meteor structure 194
Brick Building 48, 211
bridge playing 42
British Association for the Advancement of Science 17
British Columbia, Canada 25
British English xi
British Museum 111, 235
Brook Green, Hammersmith, London 108, 256
Brooke, Rupert (1887–1915), English poet 66
Browning, Robert 129
Bruce doublet 145
Bruce survey 145
Bruges, Belgium 52
Bruggencate, P. ten, German astronomer; worked in studies of stellar population 221
Brunhilda xvi
Bryant, William Cullen 178
Buccleah Mansion, New Jersey 59
Buck's Mills, England 52
Buckingham Palace, London 55
Buckinghamshire 77, 85, 129
Buckmaster, Lord 83
Building D 48
Bunbury, Sir Charles 79
Burbidge, E. Margaret (b. 1919), British-born US astronomer, spectroscopist, cosmologist xii, xv, xvi, xviii, xix
Burbidge, Geoffrey (b. 1925), British physicist xviii
Busoni, Ferruccio B. (1866–1924), Italian musician and pianist 82, 105

Caesar 156
Cal Tech (California Institute of Technology) 54
calcium, neutral 22
California 54, 183
Cambridge University, England 110, 157, 256, 257
Cambridge, England 82, 103, 115, 118, 121, 134, 155, 165, 168, 170, 219, 236
Cambridge, Massachusetts 137, 142, 145, 210, 222, 256
Cameron, A. G. W. (b. 1925), Canadian born astrophysicist xiin
camp, concentration 47
Campbell, Leon (1881–1951), US astronomer; investigated variable stars 48, 138, 152
Canada 157, 203
Canberra, Australia 64
cancer, lung 65
Cannon, Annie J., Prize xi
Cannon, Annie Jump (1863–1941), US astronomer; classified over 300 000 spectra for the *Henry Draper Catalog* and its extensions x, xiv, 7, 13, 17, 27, 29, 124, 137–40, 140p, 148, 150–1, 166p, 207, 224
Cape Cod, Massachusetts 47
Carathéodory, Constantin (1873–1944), German-born mathematician 221
card playing 42
career length xi
Caronia 130
Carrel, Alexis (1873–1944), French biologist; Nobel laureate in physiology and medicine 229
Carreño, Theresa (1853–1917), Venezuelan musician, composer, conductor 82
Carroll, Lewis (Charles Lutwidge Dodgson) (1832–1898), English writer, logician 14
Carte du Ciel 144
Cascade mountains 55
Castalian Spring, Delphi, Greece 236
Castel Gandolfo, Italy 58
Catalán, Miguel A., physicist 177

262   Index

Catherine II of Russia 78
Cattell, J. M. 23
Cavendish, Henry (1731–1810), French-born English chemist and physicist 123
Cavendish Laboratory 112, 115, 116, 159, 161
Cecchetti, Enrico (1850–1928), Italian dancer and ballet master 106
Celescope ultraviolet cameras 62
Cellini's *Perseus* 56
Center for Astrophysics 211, 258
Cepheid variables x, 122
Cepheid variables in Andromeda Nebula ix
Cepheids xiii, 170
Cepheids, Andromeda Galaxy 209
Cepheids, Magellanic xiii, 215
Chadwick, James (1891–1974), English physicist; discovered neutron 116
Chairman of the Department of Astronomy 60, 225
Chalonge, Daniel; developed three–dimensional quantitative method of spectral classification 203
Chandrasekhar, Subrahmanyan (1910–1995), Indian-born US astronomer; contributed to theoretical astrophysics, stellar evolution 203, 204p
Charing Cross, London 52
Chariots of the Gods 229
Charles River 143
Charles' Wain 86
Charmouth, England 52
chemical elements, abundance of viii, 4
chemical evolution viii
Chequers Court, Buckinghamshire 94
Chichester Canal 45
Chichester Cathedral 45
Chilterns 77
Christian Science 229
Christie, Agatha xiv, xvi
Christmas parties 48
chromodynamics 2
Chubb, Miss Sibyl 48
Cicero 102
cigarettes 43
Cimabue (*c.* 1240–1302), Italian painter 56, 57
citizen, US 256

Clarke, F. W. 4, 5, 18
classification, empirical stellar 14
classification, stellar 8
Cleopatra 156
Clerke, Agnes (1842–1907), British astronomer; author of *A popular history of astronomy in the 19th century* 30, 84
Cleveland, Ohio 17
Clouds, Magellanic x, 215, 258
Coblenz, Prussia, Rhine Province 256
Colby College 257
Colman, Ronald (1891–1958), British-born US film star xx
Color-magnitude array 178
Colosseum, Rome 58
Columbia University x, 27, 36
Comet 1847 VI x
Committee on Spectra, of International Astronomical Union 23
composition of Earth's crust 18, 19
compounds 3
Compton, Karl T. (1887–1954), US physicist 20, 23, 33
Comrie, L. J. (1893–1950), New Zealand-born astronomer 12, 29, 121, 124
Comstock, Ada 33, 34
Conant, James (1893–1978), US chemist, author, educator; President of Harvard 27, 224
concentration camp 47
Conference on Novae and White Dwarfs 203
Conklin, Nannilou Hepburn Dieter 29, 37
Copenhagen 191
Cophetua. King 156
Cornell University, New York 129
Cornwall, England 113
Council, Observatory 28
CQ Cephei 201
Cram, G. W. 37
Crater Lake, Oregon 55
Crimea 256
Crommelin, A. C. D. 34
Curator of the Astronomical Photographs 225
Curie-Joliot, Irène (1897–1942), French physicist; daughter of Pierre and Marie Curie; discovered radioactivity with her husband Frédéric 203

Curtis, Heber D. (1872–1942), US astronomer, astrophysicist; observed at 11 total solar eclipses 209
Curtis–Shapley debate ix
Cushman, Charlotte Saunders (1816–1876), US actress 141
Cushman, Florence 140p, 141, 166p
Cygnus cloud 209

Dalglish, Dorothy 43, 63, 70, 99–102
Dante, Italian poet 40
dark matter xii
Darwin, Charles (1809–1882), British naturalist; wrote *On the Origin of Species* 80, 114
Darwin, Sir Francis (1848–1925), biologist, plant physiologist 82
David of Michaelangelo 57
Daylight Comet of 1910 86, 99
Dean of Faculty (at Harvard) 27
Deering 85
Delphi, Greece 236
Demeter 41, 237
Demeter of Cnidos 235
Democritus 2
Denmark xix, 203
Department of Astronomy 157, 210
DeVorkin, D. (b. 1944), historian of science 30, 32, 33
Diaghileff, Sergei P. (1872–1929), Russian ballet impressario 106
Director's residence 211
dissertation, doctoral viii
distance scale of universe ix
doctoral dissertation viii
doctoral thesis 50
Domestic manners of the Americans 60
Dominion Astrophysical Observatory 7, 25
Draper class 8
Dresden, Germany 90
Duccio (1255?–1319?), Italian painter 56, 57
Dugan, R. S. (1878–1940), US astronomer; worked on eclipsing binaries, photometry; co-authored textbook with H. N. Russell 178
Durchmusterung chart 149
Durchmusterungen 145
Durham, England 80
Dwarf novae 216
'dyer's hand, The' vii, xix

Earth 4, 5
Earth's crust, composition of 18, 19
eclipsing binaries viii, ix
eclipsing stars 170, 198, 205
eclipsing variables 199
Eddington, Mrs 120, 121
Eddington, Sir Arthur Stanley (1882–1944), British astronomer and physicist; worked on stellar constitution, cosmology viii, xxii, 1, 12, 17, 24, 54, 117, 119, 121–4, 156, 165, 170, 182, 191, 195, 203, 204p, 230, 236
Eddington, Winifred 120
Edlén, Bengt (b. 1906), Swedish physicist; studied spectra of hot stars, solar corona 170, 204p
Edward C. Pickering Fellowship 29
Edwards, Elizabeth 70, 91, 93, 98
Egypt, visit to 64
Egyptian hieroglyphics xix
Einstein, Albert (1879–1955), German-born Swiss/US theoretical physicist; explained photo-electric effect, formulated theories of relativity 2, 51, 117, 232
elements, abundance of viii, 4
Eliot, T. S. 8
Elliott, Clark, Harvard archivist 37
Elsinore, Denmark 53
Elvey, Christian T. (b. 1899), US astro-geophysicist; worked on stellar rotation, light of night sky, aurorae 169
Emeritus Professor, Harvard University 257
Emerson, Ralph Waldo 81
empirical stellar classification 14
energy 3
England 62, 77, 79, 93, 129, 133, 175, 203, 220, 221
Ephesus, Temple of Diana 65
equilibrium, thermal viii
Estonia 192
Euclid 99
Eupatoria, Crimea, Russia 256
Euripides 102
Europe 79
evolution of galaxies xi
evolution, chemical viii
excitation, anomalous viii
extrasensory perception 229
eye for detail 7

Index

facial hair 100
family history 59
Farnsworth, Alice 36
Father O'Connell, astronomer at Vatican Observatory 58
fellowship, Edward C. Pickering 29
fellowship, Rose Sidgwick 17
Fenton, Henry J. H. (1854–1929), British chemist; worked in organic and physical chemistry 115
Field, George B. (b. 1929), US astrophysicist; worked in intergalactic matter, cosmic background radiation, galaxy formation 67, 212
'Fifty years of novae', Russell Prize lecture vii
Finland 192
first male doctoral student at Harvard College Observatory 23
First World War 235
Fitzgerald, George F. 117
flat earth 229
Fleming, Mrs. Williamina (1857–1911), Scottish-born US astronomer; discovered variables, classified stellar spectra 13, 36, 147, 151
Florence, Italy 56, 57, 80
Forum for International Problems 47, 61, 205
Forum, Rome 58
Fowler, Alfred (1868–1940), British astrophysicist; studied comets, Sun, stars spectroscopically; studied structure of spectra 17, 33, 165
Fowler, Ralph Howard (1889–1944), British mathematician; worked on thermodynamics, external ballistics, theoretical physics 7, 14, 21, 22, 31, 34, 155, 160, 161, 165
Fowler, W. A. (b. 1911), US astrophysicist, spectroscopist, studied supernovae, quasars xiin
France 79, 203
Franklin Institute 257
Fraser, Sir James, author of *The Golden Bough* 58
Fraunhofer Observatory 192
French language 40
Furness, Caroline (1869–1936), US astronomer, worked in astrometry and variable stars x, 12

galactic structure 208
galaxies 209
galaxies, evolution of xi
Galaxy 168, 170, 215
Galileo 56, 57
Gamma Velorum (star) 84, 169
Gamov, George (1904–1968), Russian-born US astrophysicist 3, 49
Gaposchkin, Edward 47, 56, 65, 222p, 256
Gaposchkin, Katherine 222p, 256
Gaposchkin, Lola Naumova (1902? – 1985) 61
Gaposchkin, Peter 44, 256
Gaposchkin, Sergei (1898–1984), Russian-born US astronomer; worked in variable stars, eclipsing binaries xvi–xx, 9, 28, 53, 84, 170, 181, 196–8, 196p, 200–1, 204p, 215, 222p, 229, 256
Gaposchkin, Vladimir (Volodya) (1907–1961) 61
Gaposchkins 213
Gaposhkin, Sergei *see* Gaposchkin
Garden Party, London 55
Garnett, Henry 83
Garnett, John (*c*. 1750–1820), British scientist 59, 79
Garnett, Julia 79
Garnetts (family) 60
Gascoigne, S. C. B., Australian astronomer; worked on clusters in our galaxy and the Magellanic Clouds 215
Gauss, Karl Friedrich (1777–1855), German mathematician, astronomer 123
Geneva, Switzerland 59
geochemistry 4
geochemists 3
geologists 3
Georgetown University, Washington D.C. 64
Gerasimovich, Boris (d. *c*. 1937), Russian astronomer; gave quantitative explanation for observed variations in certain stars 61, 180, 190, 193, 195, 197
Gerasimovich, Olga Michailovna (wife of Boris Gerasimovich) 180
German language 40
Germany 101, 191, 195, 197, 200

Germany, travels in 52
Gerrish, Professor Willard Peabody 138, 142, 157
Gervase of Tilbury (*fl.* 1212), English writer on historical subjects 232
giant planets 5
giant stars 24
Gilbert and Sullivan 8, 208
Gilchrist medal 50
Gildersleeve, Virginia 26, 36
Gill, Edith 140p, 141, 166p
Gill, Mabel 140p, 141, 166p
Gingerich, Owen (b. 1930), US astronomer, historian of science 29–31, 33, 35, 37, 50
Glarus, Switzerland xix
globular clusters ix, 168, 170, 198, 208, 215
Goethe, Johann Wolfgang von 114
Goettingen, Germany 53, 195
Gold Medal of Royal Astronomical Society xiv
Goldberg, Leo (1913–1971), US astrophysicist; investigated mass loss from cool stars, circumstellar shells, the Sun. Director of Harvard Observatory 29, 211
Goldschmidt, V. M. (1888–1947), Swiss-born Norwegian geochemist 4
Gollancz, Sir Israel (1864–1930), British scholar; authority on early English texts 83
Good King Wenceslas 48
Good Samaritan 235
Goodbye Mr. Chips xx
Gosse, Sir Edmund (1845–1928), English poet and critic, librarian to the House of Lords 230
Gozzoli, Benozzo (1420–1497), Italian Renaissance painter 57
graduate student 11
Grand Canyon, Arizona 54
Great Smoky Mountains 54
Greek language, ancient 40, 50
Green Bank, West Virginia 180
Green, Henry, Second Assistant, Cambridge Observatory, England c. 1922 119
Greenstein, Jesse L. (b. 1909), US astrophysicist x, 37, 49, 67
Greenstein, Naomi 49
Greenwich, England 220

Grenson 85
Grenson's Wood 86
Grimm, Jakob and Wilhelm, the brothers, folklorists 91, 234
Guthnick, P. 54

H-delta line 163
H-gamma line 163
Hades 41
Halley's Comet 86, 99
Hamilton, William R. (1805–1865), Irish mathematician and astronomer; invented quaternions 115
Hamlet 73
Hammersmith, London 108, 256
Hampshire, England 81
Handel, George F. 65, 105
Hanover, Germany 77, 79
happy hunting ground 258
Haramundanis, George 62, 67
Haramundanis, John 65
Haramundanis, Katherine (daughter) xiv, xxi, 37, 170
Haramundanis, Sergei 67
Harkins' rule 4
Harrow-on-the-Hill, London 96
Harvard catalogue (of courses) 25, 27
Harvard Circular No. 350 174
Harvard College Observatory 11, 29, 47, 124, 137, 144, 207, 213, 256, 258
Harvard Corporation 25
Harvard Dean of Faculty 27
Harvard Faculty Club 27
Harvard instruments 142
Harvard Observatory *see* Harvard College Observatory
Harvard photographs 177
Harvard Photometry 145
Harvard plate collection 161, 199
Harvard plates 153, 155, 201, 213
Harvard Standard Regions 145, 151
Harvard telescopes 222
Harvard University x, 85, 145–7, 149, 152, 156, 164, 180, 182–3, 197, 200, 204, 220, 224–5, 256–7
Harvey, Miss 50
Harwood, Margaret (1885–1979), US astronomer 140p, 166p
Hawaii, visit to 64
Hawkes, Julia May x
height xvii
Heinemann, Helen 37, 60

Heisenberg 2
Helsinki (Helsingfors), Finland 192
Henry Draper Catalogue x, 7, 25, 149–50, 152, 161, 168
Henry Draper classification 147
Henry Draper Extension 149
Henry Draper system 148, 161
Henry Norris Russell Prize, American Astronomical Society vii, xii, 258
Heracles 90
Herschel, Sir William (1738–1822), German-born British astronomer 120
Herstmonceaux Castle, Sussex 55
Hertzsprung, Ejnar (1873–1967), Danish astronomer; developed H–R diagram plotting luminosities of stars against their surface temperatures 84, 148, 160, 168, 170, 178, 180, 191, 209
Heyden, Francis, S. J. (d. 1993?), astronomer 64
High Wycombe, England 77, 234
Highland Farm 44, 46
Hildesheim, Germany 53
Hill, E. O. xi, xvi, xviii
Hilton, James (1900–1954), English novelist xx, 118
Hinchman, Lydia S. 29
Hippodrome, Istanbul 65
Hodgdon, Miss 140p, 141, 166p
Hoffleit, Dorrit (b. 1907), US astronomer, researched variable stars and stellar spectra 49, 51, 67
Hogg, Frank, US astronomer 23, 34, 222
Hollow Square 199
Hollywood High School, California xvi
Holst, Gustav xvi, 108
Holywell Lodge, Wendover, Buckinghamshire 256
Honolulu, Hawaii vii
Hoovens, Agnes M. 140p, 166p
Hopkins, Frederick G. (1861–1947), English biochemist; studied vitamins 114
Horbury Crescent 45
Horikoshi, Mrs 47
Horikoshi, Reverend 47
Horner, Frances 79
Horner, Joanna 79

Horner, Katherine 79
Horner, Leonard (1785–1864), Scottish geologist, social reformer, President of Geological Society of London 79
Horner, Leonora 79, 82
Horner, Mary Elizabeth, later Lady Lyell, (d. 1873), wife of Sir Charles Lyell, conchologist 79
Horner, Susan 79
Horners 43
House Unamerican Activities Committee 60
Housesteads 55
housework 42
Howe, Mary B. 140p, 166p
How Not to Do Research 164, 168–9
Hoyle, Sir Fred (b. 1915), British astronomer xiin
Hubble, Edwin (1889–1953), US astronomer; studied recession of galaxies; extragalactic region ix, 184, 209
Hufbauer, Karl, historian of science viii, 36, 37
Huggins, Sir William (1824–1910), British astronomer, spectroscopist; studied physical constitution of stars, planets, comets, nebulae 147, 150
Hull, Etheldreda 83
Humperdinck, Engelbert 120
Hund, Friedrich 221
Huxley, Thomas (1825–1895), British biologist; advocate of Darwinism 80, 98, 123
Hven, Denmark (island of Uraniborg) 53
hydrogen viii, 165
hyle 3

Icctacihuatl, Mexico 105
Iceland 81
Icelandic language 40
Index, Science Citation xiii
India 64, 161
infrared observations 23
International Astronomical Union 61, 64, 169, 190–1, 258
International Astronomical Union Committee on Spectra 23
international reputation 23
interstellar absorption 168, 171

Index 267

interstellar calcium 180
intrinsic variables 198
Introduction to astronomy 60, 61, 258
intuition 8
ionization potentials 15, 17, 21
ionization theory 21, 22
ionization, thermal 8
Isaiah 73
Ison, Leonora (b. 1904), artist, painter of architectural subjects 56
Ison, Walter, artist, authority on Georgian architecture 56
isotopes 3
Istanbul, Turkey 65
Italian language 40
Italy, travels in 52

James, Henry, US novelist 83
James, William, the Elder, US educator 81
Japan, visit to 64
Japanese–American family 47
Jason 90
Johnson, Harold Lester (b. 1921), US astronomer; developed infrared, photometric telescopes, applied electronics to astronomical problems 151
Joliot, Frédéric (1900–1958), French physicist; discovered artificial radioactivity with his wife Irène 203
Joliot-Curie, Irène; *see* Curie-Joliot, Irène
Jones, B. Z. 30
Jordan River 234
Joseph Henry Lecture 257
Josephine (heroine in *Pinafore*) 208
Joy, Alfred H. (1882–1973), US astronomer; studied stellar luminosities, distances, parallax, spectra of variable stars xi, 84
Jupiter 57, 121
'Jupiter' symphony 47

Kansas 183
Karajan, Herbert von 82
Keenan, Philip C. xi
Kenat, Ralph 30, 32
King Henry IV, Part I 63
King, A. S. 21
King, Edward Skinner (1861–1931), US astronomer; worked in photometry, astronomical photography 26, 138, 142, 145, 157, 164, 225
Kipling, Rudyard 161
Kleinmann, Susan ix, xxi
Klumpke, Dorothea Roberts (1861–1942), US astronomer, aeronaut x
Kohlschuetter, A. 34
Kopff, August, German astronomer; worked in astrometry, star positions 54, 191
Kron, Gerald (b. 1913), US astronomer; worked in astrophysics, photoelectric photometry, developed astronomical image tubes 151
Kuiper, Gerald P. (1905–1973), Dutch-born US astronomer; studied moon, planets, solar system 203, 204p

Lafayette. Marquis de 79
Lagrangian surface 216
Lake Mead, Nevada 54
Lake Nemi, Italy 58
Landscrona, Denmark 53
Lang, K. R. 35, 50
Langmuir, Irving (1881–1957), US physical chemist and inventor 116
language, native xi
Large Magellanic Cloud 215
Larson, Richard xxiii
Latin epigrams 8
Latin language 40
Latvia xix
Lauder, Harry 120
laundry 42
lava 3
Le Chatelier, Henri Louis (1850–1936), French chemist; discovered law governing how pressure and temperature affect equilibrium 115
Leaf, Betty Grierson (niece to Walter Leaf) (d. 1933) 71, 170, 190–1
Leaf, Walter (1852–1927), British banker, classical scholar 71, 170
Leavitt, Henrietta (1868–1921), US astronomer; worked on variables, cluster variables in Magellanic Clouds, found period–luminosity relation vii, x, 29, 137, 140, 145, 151, 153, 155, 170–2, 184, 198, 207, 213, 215

lecture, Russell vii, xii, 258
Lee, the, Buckinghamshire, England 94
Lehmann, I (Mrs Balanowsky), Soviet astronomer 54
Leland, Evelyn F. 140p, 166p
Lemaitre, G. (1894–1966), Belgian astronomer, originated Big Bang theory 17
Leningrad, USSR 193
Lexington Center 45
Lexington Choral Society 45
Lexington, Massachusetts 170
Leyden, Netherlands 191
Liberty, Sir Lasenby and Lady 94
Lick Observatory 7, 17, 55, 166, 183, 224
life expectancy ix
Liller, Martha (Hazen) (b. 1931), US astronomer 33, 67
Lincoln's Inn, London 96
Lindblad, Bertil (1895–1965), Swedish astronomer; helped to develop modern model of our galaxy 54, 174
Lindblads 191
'Lines Composed a Few Miles above Tintern Abbey' xixn
lines of nitrogen, analysis of 23
lines, absorption viii
Linnean system 98
Llanvihangel Court, Monmouthshire, Wales 81
Lockyer, Sir Joseph Norman (1836–1920), British astronomer; studied spectra, Ancient Egyptian astronomy 147, 159
Loggia dei Lanzi 56
London, University of xv
London, England 43, 63, 80, 94, 96, 106, 113, 158, 180
longevity record xi
Lopokova, Lydia (wife of John Maynard Keynes), ballerina 106
Lorentz, Hendrik A. (1853–1928), Dutch physicist; formulated rule called the Lorentz contraction 117
Lost Horizon xx
Louvre, Paris, France 62
low-resolution spectra 7
Lowell, A. Lawrence (1856–1943), US political scientist, educator 25, 27, 36, 222, 224
Lund, Sweden 191

Lundmark family 53
Lundmark, Knut; investigated globular clusters, galaxies 54, 180, 203, 204p, 221
lung cancer 65
Luyten, Willem I. (b. 1899), Dutch/US astronomer; worked in stellar motions, galactic structure; discovered smallest known star (1963) 13, 138, 142, 145
Lyell, Captain Thomas (brother of Sir Charles Lyell) 79
Lyell, Katherine 80
Lyell, Sir Charles (1797–1875), Scottish geologist; his *Principles of Geology* contributed significantly to nineteenth-century scientific thought 79
Lyman, Theodore (1874–1954), US physicist; worked on properties of light 31, 157, 222

M36, proper motions of stars in 12
Maanen, Adriaan van (1884–1946), Dutch-born astronomer; worked in US; studied parallaxes, proper motions of stars, nebulae xxi, 183, 209
MacDonald Observatory, Texas 180
Mack, Pamela 35, 36, 38
Madingley Road, Cambridge, England 119
Magellanic Cepheids x, 209
Magellanic Clouds x, 145, 152, 179, 198, 209, 213, 258
magnitude, absolute 22
magnitudes, ultraviolet 62
Malby, Margaret 32
Mallowan, Agatha Christie xiv, xvi
Mammoth Cave, Kentucky 54
Manchester, England xiii, 112
Marble Arch, London 96
Maria Mitchell Association 29
Marlowe, Christopher 235
marriage 63
Martineaus, Harriet and James 79
Mary Eward Scholarship for Natural Sciences 256
mass–luminosity relation 178
Massachusetts 129, 205
Massachusetts Avenue 46
Massachusetts Institute of Technology 16

masses of stars 215
Massine, Léonide (b. 1896), Russian-born US dancer and choreographer 106
Massingham, H. W. (1860–1924), British journalist, editor 94
Mater Dolorosa 237
Mathews, Francis Claughton 81
Mathews, Mary 81
matter 3
matter at high temperatures 14
matter, dark xii
matter–energy 4
Maury, Antonia (1866–1952), US astronomer; classified spectra of bright northern stars, worked on spectral changes of Beta Lyrae x, 13, 137, 140, 140p, 142, 148, 150, 160, 166p, 168, 207
Maxwell, James Clerk 31
Mayall, Margaret Walton (b. 1902), US astronomer, ran AAVSO 140p, 166p
McCarthy, Joe 60
McCrea, Sir William H. xi, xvi
measurement of individual lines, spectrophotometric 23
medal, Gilchrist 50
Medici tomb 56
memory 7
Mendel, Gregor (1822–1884), Austrian/Czech botanist 114
Menzel, Donald (1901–1976), US astrophysicist; studied the Sun, stellar spectra, planetary atmospheres, radio propagation 7, 15, 17, 21, 24, 31, 32, 34, 37, 48, 161, 164, 211, 224–5
Menzel, Mrs Florence 49
Mercury 5
Merrill, Paul W. (1887–1961), US astronomer; worked in stellar spectroscopy 34, 201
Merton, T. R. (1888–1969), English spectroscopist 201
Messiah 65, 105
Messier 3 152
Messier 8 158
metals, abundance of 18
meteorites 4
method, step 178
Mexico 46
Mexico City 105

Michaelangelo 56
Michelson, Albert A. (1852–1931), German-born US physicist; collaborated on Michelson–Morley experiment to measure ether drift 117
Michigan, University of x
Milky Way 171, 199
Milne, E.A. (1896–1950), British astrophysicist; made contributions to cosmic dynamics 7, 14, 17, 19–22, 31, 33, 34, 121, 155, 160–1, 169, 179, 221
Milton, John 178
Mineur, Henri 203
mini-clusters 215
minor planet 1974CA, Payne–Gaposchkin 258
Mira stars 172, 198–9, 215
Mitchell, Maria (1818–1889), first woman US astronomer x
Mithra 58
Mizar x
Monmouthshire, Wales 81
Montez, Lola, Countess of Landsfeld, Irish dancer; favored Liberalism 81
Monthly Notices of the Royal Astronomical Society 12
Montholon, General 59
Moore, Charlotte *see* Charlotte Sitterly
Morgan, Augustus de (1806–1871), British mathematician, astronomer, logician 81
Morgan, William W. (b. 1906), US astronomer; worked in stellar spectroscopy, galaxies xi, 50, 51, 148
Morison, Samuel Eliot 59
Morley, Edward W. (1838–1923), US chemist; collaborated on Michelson-Morley experiment to measure ether drift 117
Moscow, USSR 61, 78
Mossbauer effect, second–order xiii
Mount Chocorua, New Hampshire 178
Mount Palomar, opening of, 1948 54
Mount Wilson Observatory 7, 23, 124, 146, 150, 183, 216
Mount Wilson publications 173
Mozart, Wolfgang A. 47, 65, 105
Munch, G. xvi
Myers, Jeff 56

Index

Nantucket, Massachusetts 47
Napoleon Bonaparte 59, 178, 208
Nashoba Plan 37
National Academy of Sciences 27, 164
National Gallery, London 96
National Research Fellowship 23, 256
native language xi
naturalization xviii
Nature 238
Nature 33
Nebula, Andromeda x
Netherlands 51
neutral calcium 22
New Brunswick, New Jersey 59
New England 129, 136, 183
New Jersey 59, 79
New Mexico 54
New York, NY 64, 204
New Zealand, Wellington xviii
Newall, Hugh Frank (1857–1944), British astrophysicist; investigated solar phenomena 121
Newcastle upon Tyne 55
Newnham College, Cambridge, England xv, 82, 110, 112, 117–19, 121, 136, 256
Newnham Observatory, Cambridge, England 72
Newton's *Principia* 64
Newton, Sir Isaac (1642–1727), British scientist, philosopher; formulated theory of universal gravitation, discussed mechanics, color 2, 98
NGC 1960 122
Nijinski, Vaslav, Russian ballet dancer 106
nitrogen, analysis of the lines of 23
Niven, W. D. 31
Normandie, the 204
North Polar cap 200
North Polar Sequence 145, 151, 171, 183
North Sea 119
Northern Europe 91
Norway 181
Nova 1890, T Pxyidis vii, ix
nova, recurrent ix
novae 198, 201, 216

O stars 18
objective prism plates 7

observations in infrared light 23
observations in visible light 23
Observatory Council 28
Observatory Hill 211
Observatory Philharmonic Orchestra (OPO) 47
'Observatory Pinafore' 208
Observatory publications 174
Observatory, Harvard 156, 160, 205
O'Connell, Father, astronomer at Vatican Observatory 58
Odysseus 90
Odyssey 234
Old Vic 43, 105
Oliver, Ariadne xvi
Omega Centauri 152
Oort, Jan Hendrik (1900–1992), Dutch astronomer; studied galactic structure and dynamics; discovered rotation of galaxy with radio astronomy 191
open clusters 168
Öpik, Ernst (b. 1893), Soviet astronomer, worked on stellar distributions, meteors 51, 182, 192
Oregon 129
Orion's belt 86
Ostende, Belgium 52
Oxford Circus, London 96
Oxford, England 77, 135, 219, 234

Paddington, London 256
Palmer, Margetta (1862–1924), US astronomer x
Pannakoek, A., historian of astronomy 50
parachute 42
Paraskevopoulos, J. S. 36
Paris, France x, 56, 59, 62, 203
Pavlova, Anna, Russian ballet dancer 106
Payne (name) 77
Payne, Edward John (father) (1844–1904); British historian, Fellow of University College, Oxford 45, 77, 83, 87p, 234, 256
Payne, Humfry (1902–1936), British archaeologist, Director of British School of Archeology, Athens; discovered Perachora, joined Rampin Head fragments 53, 86, 94
Payne, Leonora (b. 1904), British

architect, artist; writer on architectural subjects 86
Payne, Miss 88p, 130p, 135p, 138p, 140p, 166p, 185p, 223
Payne, Mrs Edward 16, 32, 53
Payne-Gaposchkin Principle 162
Payne-Gaposchkin, Cecilia (photos) 204p, 214p, 222p, 224p, 226p, 257p
Paynes 45
Pearl Harbor 181
Pease, Francis G. (1881–1938), US astronomer; designed 100-inch telescope, method for grinding 200-inch reflector 183
Pegram, George G. 36
Pendlebury, Ivy 70, 109, 114
period-luminosity relation x, 170, 178
Persephone 41
Perseus 90
Perseus twin clusters 183
Perseus, Cellini's 56
Pertzs 45
Pertz, Annie (1855–1920?), painter 82
Pertz, Christian August 78
Pertz, Dorothea (Dora) (d. 1939), botanist 43, 82, 113
Pertz, Eddie (b. 1863) 82
Pertz, Elizabeth (1857–1862) 82
Pertz, Emma (mother) (1866–1947) 78 (portrait), 82, 256
Pertz, Florence (1872–1956?) 55, 82
Pertz, Fritz (b. 1860) 82
Pertz, Georg Heinrich (1795–1876), German historian, member of parliament 45, 59, 79, 82
Pertz, George 83
Pertz, Hermann (1832–1879/80), Prussian engineer 79, 81–2
Pertz, Julia Garnett (1793–1852), great-grandmother 59
Pertz, Karl 83
Peru 145, 207, 210
Pettit, Edison (b. 1890), US astronomer; investigated Sun, biology, ultraviolet light 183
Petty Cury, Cambridge, England 72
Phelps, William Lyon (1865–1943), US educator and literary critic 133
Phi Beta Kappa 258
Philadelphia, Penna. xii
Philippines 64

Phillips Astronomer 225, 256
photographic photometry viii, 26
Piccadilly Circus, London 96
Pickering, Edward C. (1846–1919), US astronomer; Director Harvard Observatory, worked in stellar photometry, classification of spectra 137, 142, 144–5, 147–8, 151–2, 154, 156, 171, 202, 207, 208, 210
Pierce, Newton, US physicist 178
Pisa, Italy 56
Pitti Palace 56
Pitti, Luca 56
planets, giant 5
Plaskett, Harry H. (1893–1980), Canadian astronomer; studied solar, stellar spectra, solar granulation, Sun's differential rotation 18, 32, 37, 157, 183, 223–4
Plaskett, John S. (1865–1941), Canadian astronomer; Director Dominion Observatory, Victoria; spectroscopist, found largest known star (Plaskett's) 17–19, 25, 32, 35, 169, 223
Plaskett, Mrs 223
plate library 160
plates, objective prism 7
Plato 73, 102, 110, 219
Plummer, Henry C.; worked on variable stars 122
Poincaré, Jules Henri (1854–1912), French mathematician, physicist, philosopher 119, 230
Poirot, Hercule xiv
Pope, the 178
Popocatepetl, Mexico 105
populations, stellar viii
potential novae 216
Powell, Dilys (1901–1995), film critic, wife of Humfry Payne 53
Prager, Richard, German astronomer; worked on variable stars 46, 200–1
pre-Socratic model 2
Preston, Jean 37
Priestley, Joseph (1733–1804), British chemist, clergyman; discovered oxygen 79
Princeton, New Jersey 161
Principe (island of solar eclipse) 12
prize, Annie J. Cannon xi, 256

Index

prize, Henry Norris Russell vii, xii, 258
Proceedings of the National Academy of Sciences 20
professor 28, 60
Professorship 211
Prometheus 237
proper motions in M36 12
Ptolemaic system 114
Pulkova, USSR 53, 61, 180, 191, 193, 195
Punch 73
Pythagoreans 2

quantum electrodynamics 2
quantum mechanics 2

R Corona Borealis variability xiii
Radcliffe College, Cambridge, Massachusetts x, 16, 24, 25, 111, 135, 157, 256–7
Radcliffe, no women on faculty of 25
Ramanujan xx
Random Harvest xx
Ranier, Charles xx
Rankine, Professor 206
Reade, Charles 179
reasonableness 43
recurrent nova ix
recurrent novae 201
Reeth, Swaledale 55, 62
relative abundance of chemical elements viii, 4, 6
reputation, international 23
Reval (Tallinn, Estonia) 53, 192, 195
Rhijn, P. J. van 32
Richard III 73
Riga, Latvia 195
Rittenhouse Medal, Franklin Institute 257
Riviera 84
Roberts, Dorothea *(see* Klumpke) x
Rome, Italy 57, 58
Romeo and Juliet xxin
Rose Sidgwick Fellowship 17, 32
Rosseland, Svein (b. 1894), Norwegian astrophysicist; worked in theoretical astrophysics, pulsation theory of variable stars 24, 35, 180
Rossiter, Margaret, historian of science 35
rotation of Galaxy 174
Rowland's data 18
Rowland's tables 23, 34
Royal Astronomical Society, Monthly Notices of 12
Royal Astronomical Society, London xiv, 12, 180, 209, 258
Royal Observatory 220
RR Lyrae stars 152, 170, 198, 215
RU Lupi 8
Rubin, Robert xviii
Rubin, Vera, (b. 1928), US astronomer; discovered evidence of dark matter xiin, xix, 37, 67
Rumker, Mme, co-discoverer of Comet 1847 VI x
Russell Lecture vii, xii, 63
Russell, Henry Norris (1877–1957), US astronomer; worked in stellar evolution, contributed to H–R diagram vii, viii, 4, 6, 14–27, 30–3, 35–7, 49, 51, 60, 120, 150, 160–1, 163, 169, 177–8, 182, 203, 204p, 207–8
Russia 61, 77, 129, 181, 192, 195, 197, 256 *(see also* USSR)
Russian language 40
Russian War Relief 61
Rutherford, Eileen 118
Rutherford, Ernest (1871–1937), British physicist; suggested existence of atomic nucleus 2, 112, 116–18, 191
RY Scuti (variable star) 84, 201
Saha, Meghnad N. (1893–1956), Indian astrophysicist; derived Saha equation relating ionization rates with temperature, electron pressure 7, 14, 19–22, 28, 30, 51, 155, 160, 179
salary 36
  1930 26
Salpeter, Edwin E. (b. 1924), Austrian–born US astrophysicist; worked in quantum theory, quantum electrodynamics, theoretical astrophysics 182
Samarkand, Tashkent, USSR 61
San Clemente, Rome, Italy 58
San Gimignano, Italy 57
Santa Maria sopra Minerva, Rome, Italy 58
Saturn 55, 121
Saunders. F. A., physicist 16, 17, 31, 32, 37

Savior 237
Sawyer, Helen, Hogg (1905–1993), US astronomer x
Schaler, A., organizer of Paris 1939 conference 204p
Schlesinger, Frank (1871–1943), US astronomer; worked on stellar distances, solar rotation, astrometry 174
scholarship, Mary Eward 256
Schumann, Robert, musician 82
Schwarzchild, Karl (1873–1916), German astronomer; worked on comets, preferential stellar motions, stellar atmospheres 151
Schwarzschild, Agathe 54
Schwarzschild, Barbara 49
Schwarzschild, Martin (b. 1912), German-born US astrophysicist 49
Schwarzschilds, family 49
science and mythology 228
Science Citation Index xiii
scientometrics xiii
Scopas 236
Scotland 62
Scott, Dunkinfield Henry (1854–1934), British botanist; worked on extinct plants, plant evolution 112
Scutum Cloud 209
Seares, Frederick H. (1873–1964), US astronomer; worked on theory of orbits, solar magnetic field, stellar distributions 146, 151, 172–3, 183
Searle, George F. C. (1864–1954), British scientist; worked in electromagnetism, measurement of magnetic hysteresis 116
Sears, R. L. (b. 1927), US astrophysicist, photometrist x
Secchi, Pietro Angelo (1818–1878), Italian astronomer; first to classify stars by their spectra (color) 147
second-order Mossbauer effect xiii
Sergei xvii, xx
Seuss, H. E. 4
Seward, Albert C. (1863–1941), British paleobotanist 113
Seznec, Jean 231
Shade Street 45, 47

Shakespeare, William xxi, 10, 63, 105–7, 156
Shapley, Harlow (1885–1972), US astronomer; Director Harvard College Observatory, worked in photometry, spectroscopy, cosmology vii, ix, xxi, 12, 15–17, 22, 23, 29–34, 36, 37, 48, 50, 60, 124, 137, 139, 142, 146, 150–1, 153–7, 161–3, 165–6, 168–71, 173–4, 176–8, 183–4, 189–90, 198–9, 205, 207–8, 210, 213, 216, 221, 228–9
Shapley, Mrs Martha (Betz) (1890–1981) 48
Shaw, George Bernard, playwright 133, 156
Siena, Italy 57
Sigma Xi 257, 258
Sismondi, Jean Simonde de (1773–1842), Swiss historian 79
Sitter, Willem de (1872–1934), Dutch astronomer, cosmologist 84
Sitterly, Banny, US physicist 49
Sitterly, Charlotte Moore (1898–1990), US astrophysicist; analyzed atomic spectra, compiled tables of multiplets, atomic energy levels xii, 24, 32, 35, 37, 49, 67, 178
Skåne, Sweden 53
sky survey, two-micron ix
Small Magellanic Cloud variables xiii
Smart, William Marshall (b. 1889), British astronomer, author; worked in celestial mechanics, stellar kinematics, combination of observations 12, 119–20, 122
Smith College 257
Smith, Elske v. P., US astronomer 28, 37, 51
Smithsonian Astrophysical Observatory 49, 211, 225, 258
solar abundance of hydrogen and helium 20
solar spectrum 24
Sophocles 102
South Africa 48, 190, 210
  Boyden station 25
South America, tour of southern observatories 64
southern observatories, tour of 64
Soviet Union 53, 61, 193

Spain, visit to 64
spectra, low-resolution 7
spectra, stellar 22
spectrophotometric measurement of individual lines 23
spectroscopic binary Mizar x
spectrum, solar 24
spectrum, variable of 8
Spencer Jones, Sir Harold (1890–1960), Astronomer Royal 55
spice shelf 41
Spinoza, Benedict de 114
spiral nebulae, nature of ix
SS Batory 55
SS Cygni 141
St. Gothard pass 56
St. John, Charles Edward 34, 183
St. Mary's College, Paddington 256
St. Matthew 110
St. Paul's Girls' School, Brook Green, Hammersmith, London 108, 256
St. Peter's basilica, the Vatican 58
Stalin, Joseph 195
standard photometry 171
Standard Regions 145
Stanley, Wendell (1904–1971), US biochemist; contributed to understanding of chemical nature of viruses 115
star clusters 216
Star clusters 174, 216
Stark Effect 169
Stark, Johannes (1874–1957), German physicist; discovered Stark Effect (splitting of spectrum lines with light source in a strong electrostatic field) 121
stars 5
Stars and clusters ix, 258
Stars in the making 58, 60, 63, 258
stars of high luminosity 198
stars, binary, in clusters ix
stars, giant 24
Stebbins, Joel (1878–1966), US astronomer; worked in photometry of stars and interstellar space xi, 151
Stebbins, Joel (1878–1966), US astronomer, worked in photometry of stars and interstellar space xi, 151
Stellar atmospheres 5, 6, 9, 11, 20, 23–4, 28, 33–4, 165, 168, 256

stellar atmospheres, abundance of elements in 19
stellar atmospheres, temperature scale for 21
stellar classification 8
stellar populations viii
stellar pulsation 170
stellar spectra 22
step method 178
Stetson, Harlan 16
Stewart, J. Q. 34
Stockholm, Sweden 191
Stratton, Frederick J. M. (1881–1960), British astrophysicist; worked in solar, stellar physics 120, 191, 203, 204p
Strömgren, Bengt (1908–1987), Swedish astronomer; worked in astronomy, experimental physics 25, 35, 191, 203, 204p
Strömgren, Elis (1870–1947), Swedish astronomer; studied motions of double stars, comets 191
Struve, Friedrich Georg Wilhelm von (1793–1864), German-born Russian astronomer; first measured parallax of Vega 192
Struve, Otto (1897–1963), Russian-born US astronomer; discovered interstellar matter xiv, 20, 33, 50
student, graduate 11
Sun 5, 160
Sun's atmosphere 18
Sunday School, Unitarian Church 45
Supergalaxy Hypothesis 174
supergiants 9
supernovae xiv, 180, 198
Swarthmore College, Pennsylvania 13, 124
Sweden 119, 129, 181, 192, 203
Swedenborg, Emanuel (Svedberg) (1688–1772), Swedish scientist, theologian 81, 98, 167, 235
Swings, Pol F. (b. 1906), Belgian astrophysicist, studied spectra of comets 204p
Swope, Henrietta (1902–1980), US astronomer; worked in photometry, variable stars 49, 170, 199

T Ceti (variable star) 84
T Pxyidis, nova 1890 vii, ix
Tahiti 64

Tashkent, USSR 61
Tate Gallery, London 96
telephone 42
telescope, 15-inch 48
telescope, Boyden 33
temperature scale for stellar atmospheres 21
Temple of Diana at Ephesus 65
Tennyson, poet xiv
Tewksbury, Massachusetts 66
Thanksgiving 1941 46
'The dyer's hand' xixn
The galactic novae vii, 50, 60, 258
The Garnett letters 59
The hunting of the snark 73
The magic flute 65
'The Skipper' – restaurant in Nantucket 47
The stars of high luminosity 8, 9, 168
Theophrastus (fourth–third centuries BC), Greek philosopher, botanist 114
theory of atom, Bohr's 21
thermal ionization 8
thesis advisor viii
thesis, doctoral 50
thin wedge 29
Thirkill, Henry, British physicist 220
Thomas, Gladys 49
Thomas, Richard N. (b. 1921), US astrophysicist 49, 50
Thompson, J. J. (1856–1940), British physicist; demonstrated existence of electron 115, 116
Through the Looking Glass 14
Tikhoff, Soviet spectroscopist 53
Tinsley, Beatrice Muriel Hill (1941–1981), New Zealand-born astronomer, worked on evolution of galaxies xi, xiin, xv, xvii, xix
Tinsley, Brian xviii
Titchener, Edward Bradford (1867–1927), English psychologist, follower of Wundt 129
Tonanzintla Observatory, Mexico 46
Toronto, Canada 17
travels in Austria 52
travels in Germany 52
travels in Italy 52
Trinidad and Tobago, visit to 64
Trinity College, Cambridge, England 117, 120
Tripos 220

Trollope, Anthony 60
Trollope, Fanny (1779–1863), English novelist; mother of Anthony Trollope 60, 79
Trumpler, Robert J. (1886–1956), Swiss-born US astronomer; investigated galactic star clusters, interstellar scattering 10, 168, 184
Tufts Medical School 66
Turku, Finland (Åbo) 192
Turner's *Chichester Canal* 45
Turner, Herbert H., (1861–1930),British astronomer; worked in photometry 220, 223
Turner, Joseph M. W. (1775–1851), English landscape painter 96
two-micron sky survey ix

U Geminorum stars 198
U Gruis (variable star) 84
U Scorpii 201
US Geological Survey 4
Uffizi Gallery, Italy 56, 105
UFO 229, 232
ultraviolet magnitudes 62
Underhill, Anne (b. 1920), Canadian-born US astrophysicist xi, 50
unification theory 2
Unitarian Church Sunday School 45
United States 77, 81, 111, 124, 129, 132, 134, 137, 157, 167, 179, 181, 183, 193, 203, 220
universe distance scale ix
University College, Oxford, England 234
University of Michigan x
University of Texas at Austin xi
University Printing Office 173
University, Harvard (*see also* Harvard University) 222
Unsöld, Albrecht (b. 1905), German astrophysicist; worked in stellar atmospheres 24, 35, 179, 221
upsilon Sagitarii 9, 140, 201
Uraniborg, observatory of Tycho Brahe 53
Urey, H. C. (1893–1981), US chemist, received Nobel Prize for Chemistry for discovery of heavy water and deuterium 4
US citizen 256
USSR (*see also* Russia, Soviet Union) 193

UW Canis Majoris 201
V2 rocket 54
V2000 Sagitarii 152
V444 Cygni 201
V453 Scorpii 201
van Maanen (*see* Maanen, van)
variability of R Corona Borealis xiii
variable of spectrum 8
variable stars viii, xiv, 170–1, 199–200, 215
Variable stars 199
Variable stars and galactic structure 60
variables in Small Magellanic Cloud xiii
Vassar College 29, 142
Vatican Museum 58
Vatican Observatory 58
Velikovsky 58
Venus de Milo 37
Verster, Floris Hendrik (1861–1927), Dutch painter 191
Victoria, British Columbia, Canada 55, 157, 223
Visher, S. S. 34
visible light, observations in 23
Vogel, Hermann Karl (1841–1907), German astronomer; discovered spectroscopic binaries 147
VV Cephei 201
VV Orionis 178

Walker, Arville D. (Billy) (d. 1963), US astronomer 29, 140p, 142, 166p
Walker, Merle F. (b. 1926), US astronomer; worked in photoelectric photometry, electronic image intensification 84
Walton, Margaret L. (*see* Mayall, Margaret)
Ward, Geneviève (1838–1922), US actress, played in Britain, worldwide 81, 106
Warner, Debby J. 35
Washington, DC 197
Washington, H. S. 4, 5
Waterfield, Bill 27, 190
Weber, Joe xviii
wedge, thin 29
Weizaecker, Karl F., Baron von (b. 1912), German astronomer, cosmologist, physicist 182
Wellesley College, Massachusetts 133

Wellington, New Zealand xvii
Wells variable 141
Wells, Louisa D., US astronomer, worked in variable stars 141
Welther, Barbara, US historian of science 67
Wendover, England 77, 94, 97, 114, 234, 256
Whipple, Babby 49
Whipple, Fred L. (b. 1906), US astronomer, inventor, Director Smithsonian Astrophysical Observatory; contributed to theory of comets, solar system 29, 46, 49, 51, 211
White Sands Proving Ground, New Mexico 54
Whitehall, London 204
Whitney, Charles A. (b. 1929), US astronomer 29, 37, 67
Wickson, Gladys 27, 36, 37
Wiener, Norbert (1894–1964), US mathematician 32
Wilkinson, Emma Marsh (1843?–1893) 80, 82
Wilkinson, Florence 81
Wilkinson, James John Garth (1812–1899), physician 80–1, 83, 92
Wilkinson, John, ironmaster 80
Wilkinson, Mary (1847–1944) 81
Wilkinsons 45
Wilson College 257
Wilson, Charles Thomson Rees (1869–1959), British physicist; devised the cloud chamber 115
Wilson, Harvia Hastings 29, 138p, 140p, 166p
Wilson, Woodrow 112
Wilstaetter, Richard (1872–1942), German chemist; won Nobel prize for research on plant pigments 115
Wimbledon Church, England 113
Wisconsin 45
Wolf–Rayet stars 169
woman years x
women on faculty of Radcliffe 25
Women's Medical College, Philedelphia 257
women, appointments from Harvard 25
Wood, Alexander 116, 230
Woods, Ida E. 140p, 141,166p

Index

Wordsworth xixn
World War II 61
Worlds in collision 58
Wright, Camilla, sister of Fanny Wright 37
Wright, Fanny (Frances D'Arusmont) (1795–1852), Scottish free-thinker and abolitionist 37, 79
Wright, Frances (1897–1989), US astronomer; worked in variable stars, meteors, photometry, navigation, much beloved teacher 32, 37, 47, 49, 183
WX Ceti xviii, 201

Yale Catalogue of Bright Stars 145
Yale University x

Yalta, Crimea, USSR 61
Yerkes Observatory, Wisconsin 45, 180
ylem 3
Yosemite Valley, California 55
Young, Judith, US astronomer xiv, xxi

Z Centauri 85
Zebergs, Velta 20, 33
Zeeman Effect 169
Zion Canyon 54
Zwicky, Fritz (1898–1974), Swiss astronomer, studied supernovae, clusters of galaxies xiv, xviii

Printed in the United States
73648LV00002B/156